THE STATE, TECHNOLOGY AND INDUSTRIALIZATION IN AFRICA

Also by Jeffrey James

CONSUMER CHOICE IN THE THIRD WORLD
CONSUMPTION AND DEVELOPMENT
ECONOMICS OF NEW TECHNOLOGY IN DEVELOPING COUNTRIES (*with F. Stewart*)
IMPROVING TRADITIONAL RURAL TECHNOLOGIES
TECHNOLOGY, INSTITUTIONS AND GOVERNMENT POLICIES (*with S. Watanabe*)
THE IMPACT OF WORLD EMPLOYMENT PROGRAM RESEARCH ON TECHNOLOGY (*with F. Stewart, A. Bhalla and others*)
THE TECHNOLOGICAL BEHAVIOUR OF PUBLIC ENTERPRISES IN DEVELOPING COUNTRIES
THE TRANSITION TO EGALITARIAN DEVELOPMENT (*with K. Griffin*)

The State, Technology and Industrialization in Africa

Jeffrey James
Professor of Development Economics
Director of the Center Graduate Schools in Economics and Management
Tilburg University, The Netherlands

M
St. Martin's Press

First published in Great Britain 1995 by
MACMILLAN PRESS LTD
Houndmills, Basingstoke, Hampshire RG21 2XS
and London
Companies and representatives
throughout the world

A catalogue record for this book is available
from the British Library.

ISBN 0–333–59557–2

10 9 8 7 6 5 4 3 2 1
04 03 02 01 00 99 98 97 96 95

Printed in Great Britain by
Ipswich Book Co Ltd
Ipswich, Suffolk

First published in the United States of America 1995 by
Scholarly and Reference Division,
ST. MARTIN'S PRESS, INC.,
175 Fifth Avenue,
New York, N.Y. 10010

ISBN 0–312–12491–0

Library of Congress Cataloging-in-Publication Data
James, Jeffrey.
The state, technology and industrialization in Africa / Jeffrey James.
p. cm.
Includes index.
ISBN 0–312–12491–0
1. Technology and state—Africa. 2. Industrialization—Africa.
I. Title.
T28.A1J36 1995
338'.06—dc20 94–31256
CIP

To the memory of Maurice Ross James

Contents

List of Table

List of Tables

Acknowledgements

Part of the research on which this book is based began while I was a staff member of the Technology and Employment Branch of the International Labour Office, Geneva. That research formed part of a larger project on public enterprises in a range of different developing countries and was published in a book under my editorship entitled *The Technological Behaviour of Public Enterprises in Developing Countries* (Routledge, 1989). In a slightly amended (and retitled) version, Chapter 3 is reproduced from an article in *Macro-Policies for Appropriate Technology in Developing Countries*, edited by Frances Stewart (Westview, 1987). I am grateful to Professor Stewart for allowing me to reproduce that article. Parts of Chapters 1, 4 and 8 draw on a paper prepared in 1989 for the Africa Region of the World Bank, entitled 'the Institutional Environment for Industrial Technology in Sub-Saharan Africa: A Study of Large-Scale Investment Projects'.

The author and publishers wish to thank the following who have kindly given permission for the use of copyright material: Longman Information and Reference for Table 3.3 from J. Forge, *Science and Technology in Africa*, copyright 1989; Lynne Rienner Publishers for Table 8.5 from *Africa's Development Challenges and the World Bank: Hard Questions, Costly Choices*, edited by Stephen K. Commins, copyright © 1988 by Lynne Rienner Publishers. Used with permission of the publisher, the OECD for table on p. 183 of DAC, *Development Co-operation*, OECD, Paris, 1986; Elsevier Science BV for Table 3 from H. Pack, 'Aggregate Implications of Factor Substitution in Developing Countries', *Journal of Development Economics*, vol. 11, 1982 and for Table 1 from M. Roemer, G. Tidrick and D. Williams, 'The Range of Strategic Choice in Tanzanian Industry', *Journal of Development Economics*, vol. 3, 1976; Oxford University Press for table on p. 287 of *Tanzania: A Political Economy*, 1982, by permission of the publisher; The World Bank for Table 16 in W. Steel and J. Evans, *Industrialization in Sub-Saharan Africa*, Technical Paper No. 25, 1984, Table 2.1 in *Financing Adjustment with Growth in Sub-Saharan Africa*, 1986 and table on p. 102 of *World Development Report*, 1983; International Monetary Fund for table on p. 2 of P. Heller and A. Tait, 'Government Employment and Pay: Some International Comparisons', IMF Occasional Paper No. 62, Washington, DC, 1984; James Currey Limited and the editors for table on p. 249 of 'The Role of Government Institutions in Kenya's Industrialization' by G. Ikiara in *Kenya's*

Industrialization Dilemma edited by P. Coughlin and G. Ikiara, Heinemann and James Currey, 1991; Prof. S. Wangwe for table on the growth of industrial capacity in Tanzania from his paper 'Industrialization in Tanzania', mimeo, 1983; Rwekaza Mukandala for Table 48 on p. 316 of his thesis on the 'Political Economy of Parastatal Enterprise in Tanzania and Botswana', University of California, Berkeley, 1988.

I am grateful to T.M. Farmiloe of Macmillan for allowing me several extensions in the delivery date of the final typescript. Corina Maas typed the many revisions of the draft chapters with considerable patience and good humour, while Anne Rafique edited the volume skilfully and thoroughly.

Introduction

If for no other reason than the sheer extent of its involvement, the role of the state is difficult to ignore in any study of technology and industrialization in Africa. To a far greater extent than elsewhere in the Third World, the industrial sector in Africa is dominated by public sector institutions[1] and to a correspondingly greater extent than in other regions, therefore the technological aspects of industrialization in African countries have to be understood in relation to the behaviour of those and other state institutions. Yet, while a considerable amount is known about how such institutions have behaved, we have only a very limited idea of *why* they have actually done so.

To a large extent this discrepancy in our knowledge reflects the emphasis of the literature – comprising as it does, mostly (micro) economics contributions – on outcomes rather than on the procedures that lie behind them.[2] Such comparative neglect of procedural issues would, of course, matter less if observed outcomes were thought to have promoted industrialization in Africa, or, even if this was not thought to be the case, if observed behaviour could be readily explained in terms of conventional theory. It is our contention, however, that neither of these propositions accurately describes the African experience during the 1970s and early 1980s; we shall argue instead, that the technological problems of the public sector have contributed in large measure to the problems now confronting most countries on the continent and that over quite a wide range of this sector's behaviour a number of distinctly paradoxical features can be observed.

Perhaps the most intriguing of these is the apparent inconsistency in the technological choices made by public enterprises *within* particular countries. Whereas, that is, one would reasonably expect the same state to make similar types of technological choices (under similar conditions), in practice just the opposite has sometimes occurred. Research on the textile industry in Tanzania, for example, has revealed a very wide range of factor proportions among state-owned enterprises, which cannot be explained with reference to most usual micro-economic models of the choice of technology.[3] In the manufacture of bricks in that same country, at the same point in time, 'Two very different types of technological choice have been made by the "same" state, one of an almost unbelievably inappropriate nature, the other much more relevant.'[4] How, then, one might ask, is it possible for the state simultaneously to maintain such technologically divergent preferences?

Comparisons *between* African countries raise a number of different, though no less puzzling, questions. Case-studies of large numbers of public enterprises in countries as diverse as Kenya, Ghana, Zambia and Tanzania, for example, reveal very *similar* patterns of technological behaviour when one might reasonably have expected certain important differences to manifest themselves. One might have expected, for example, 'that as a socialist country with a strong commitment to an egalitarian approach to development, the political pressures which have elsewhere contributed to the encouragement of inappropriately capital-intensive technologies would have been almost nonexistent'.[5] Yet, no less than in other African countries, the overly capital- and import-intensive technologies chosen by the majority of public enterprises in Tanzania have (as Chapter 1 all too clearly shows) contributed significantly to the overall efficiency problem in the public sector and have made no more than the extremely limited contribution to other major economic and social objectives (such as growth, employment and regional equality) that typifies the public enterprise in African industry.[6]

One might also have expected to find that, in countries such as Tanzania, the public sector had been explicitly used to promote the acquisition of at least the most fundamental indigenous technological capabilities, for self-reliance was, after all, one of the major guiding socialist principles proclaimed by the leaders of that country in the period after independence. Yet it seems no less true of Tanzania than other African countries that technological capabilities – of even these most fundamental kinds – have been systematically neglected by those responsible for the implementation of development projects in the industrial sector.

One has to wonder too how such disappointing outcomes could have occurred in conditions where investment projects were in principle subject to elaborate scrutiny not only by the national planning institutions of the countries concerned, but also by the numerous aid donors who (as shown in Chapter 1) have been so prevalent in African industry. Among the latter, the World Bank perhaps merits particular comment from this point of view, not only because of the stringency of its project appraisal methods but also because of its proclaimed commitment to the very types of technologies that public enterprises have mostly chosen to eschew. In particular,

> The Bank's techniques for preparing and appraising projects are intended to ensure that the technology used is tailored to the social, economic, cultural, and environmental needs of a specific situation[7] . . . the

> Bank encourages whenever possible the use of technologies which will minimize the capital costs per job created[8] . . . Preparation of a project for World Bank financing must incorporate consideration of the full range of technical, economic, financial, and institutional conditions necessary to achieve the project's objectives.[9]

Generally speaking, the method of analysis that is used below to explain these various paradoxical aspects of industrialization in Sub-Saharan Africa is probably best described as 'Hirschmanian'. For, on the basis of a large amount of detailed case-study evidence and by drawing on a wide range of literature from several different disciplines, what this book essentially describes is a search for the various 'hidden rationalities' that in combination underlie the observed patterns of technological and other forms of economic behaviour in the public sector. As such, the analysis bears a rather close methodological resemblance to the work of Albert Hirschman, who, in various contexts and at different times, has been able to reveal the 'hidden rationality of seemingly odd, irrational, or reprehensible . . . behavior'.[10] (Indeed, so central has been this concern to his many important contributions to development economics, that he himself describes it as providing an 'underlying unity' to his work.)[11]

THE BUREAUCRATIC CHOICE OF TECHNIQUES

Almost all the chapters in the first part of the book are based on, or are related to, an article about the textile industry in Tanzania that was written almost twenty years ago by David Williams with the intention of explaining why state-owned firms in that industry seemed to choose such highly disparate types of technology. In addressing this question, what the article described was a particular decision-making process in the public sector, which, in spite of being referred to as the 'bureaucratic choice of techniques', in fact exhibits a very striking *inattention* to technological factors of all kinds.[12]

This somewhat paradoxical feature of the process reflects, among other things, the nature of the objective function that the decision-maker in the public sector is thought to maximize. In contrast to almost all the choice of technology literature which imputes *technological* preferences of one kind or another to individual decision-makers,[13] the preference function of the bureaucrat is said to contain few if any such technological proclivities. His goals are expressed rather in terms of projects and in particular the quantity of projects that he is able to initiate and implement.

Since national planners in Tanzania during the 1970s were most receptive to projects which 'brought their own foreign exchange', the bureaucratic concern with initiating projects came, so the argument goes, to be translated into the goal of securing external finance. And since the technology for investment projects is typically closely linked to the financial source (be it an aid donor or an export credit guarantor), it follows that 'the "choice of technique" in any particular project is often merely a fallout from the chosen source of finance and related project inputs'.[14] What Williams refers to as the 'bureaucratic man' hypothesis may then be expressed by the notion that 'the bureaucratic decision-maker attempts to achieve objectives which *eliminate* the choice of technology from the decision process'.[15] Far, therefore, from resembling the purposive search for and the comparison of alternative technologies that is suggested by traditional micro-economic theory, what this hypothesis depicts is behaviour at just the opposite extreme, where technological issues receive little or no attention.

Chapter 2 shows how this approach was used originally to explain the seemingly paradoxical choices of technology made by different public enterprises in Tanzania's textile industry and it also indicates why those choices could not be accounted for by more conventional approaches to the question. Chapter 3 argues that the unexpected degree of similarity in the behaviour of public enterprises in (capitalist) Kenya and (socialist) Tanzania, derives to an important extent from the same bureaucratic preoccupation in both countries with foreign exchange maximization, which, in turn, led to a heavy reliance on Western sources of finance and technology. These technologies, as one might expect, were frequently large-scale, capital-intensive and inefficient (in the African context) in a variety of different senses of the term. Chapter 4 is also based on a comparison that involves two different African countries, Kenya and Zimbabwe; unlike Chapter 3, however, the objective is not to show how the bureaucratic choice of techniques can be used to explain the similar patterns of technological behaviour in two countries whose ideological differences might have been expected to produce dissimilar patterns. The comparison (in Chapter 4) is rather between the bureaucratic choice of techniques (exhibited in the Kenyan case) with another type of decision-making that (in the example from Zimbabwe) corresponds much more closely to the behaviour of 'economic man'. In Chapter 5 the same historical perspective that was used in previous chapters to explain the bias in favour of large-scale, capital-intensive technologies in the public sector, is used (in conjunction with a decision-tree type of analysis) to understand some of the exceptions to this general pattern of technological behaviour in Kenya and Tanzania.

What is common to most of the examples of appropriate technology that are described in Chapter 5 is a close connection – in one form or another – to developing rather than developed country sources of technology and technical knowledge. Chapter 6 extends the notion of a 'bureaucratic choice of techniques' to cover aid donor institutions; the argument there is that the motivation, and hence much of the behaviour of those institutions, can be understood in terms that are very similar to the patterns of decision-making already described in relation to public sector institutions in several recipient countries.

If these five chapters are thus able to throw light on certain important aspects of technology and industrialization in Africa, they also raise a further set of questions. In particular there is a rather striking lack of *political* analysis of the bureaucracy in Part I, in spite of the fact that '*The growth of the public bureaucracy in African states has been one of the most notable formal features of African politics in the past three decades.*'[16] Indeed, according to some observers 'Civil servants and government employees . . . emerged as the core of a new dominant class in the post-colonial period. This state managerial stratum – often called a bureaucratic bourgeoisie – derived its strength (and its wealth) from its proximity to state resources. The shift toward statism made politics a main avenue to material advancement and hence a key criterion for social differentiation'.[17]

Political scientists have been at pains to examine the complex and varied ways in which this political dominance of the bureaucracy has influenced the post-colonial African state. And though it is not our intention to review the entire literature that these research efforts have produced, there are a number of quite specific respects in which a political economy analysis of the bureaucracy illuminates the processes that were described in Part I. It is accordingly to an exploration of these political dimensions of the relationship between the bureaucracy and technology that the second part of the book is devoted.

THE POLITICAL ECONOMY OF BUREAUCRACY

Defined very generally as the extension of economics to the political realm, the 'new political economy', or, as it is sometimes called, the 'public choice' approach, has already been applied in a number of different ways to the African context. A well-known study by Bates, for example, sought to explain 'the political calculations that induce governments to intervene in ways which are harmful to the interests of most farmers'.[18] More recently,

Gallagher has examined the connections between rent-seeking behaviour and economic growth across a wide range of African countries.[19] What has not yet been applied to these, or indeed, as far as we are aware, to any other developing countries, is the area of public choice theory that deals with the goals of bureaucracy.

Yet, as Chapter 7 argues, this branch of the public choice literature is able to furnish much of the theoretical underpinning currently lacking in what was referred to in Part I as the 'bureaucratic choice of techniques'. What is most lacking in this latter approach is an understanding of the motivational foundations of bureaucratic behaviour at the micro-economic level. Though the question of bureaucratic objectives is partially addressed in Part I, it is certainly not discussed at any great length there and despite its centrality to the argument of that part, remains far from being well understood. In particular, while it is understandable that profitability has not been a major bureaucratic concern in African countries that, until fairly recently, were pursuing a highly inward-looking pattern of industrialization, it is much less clear exactly how and why other objectives – especially preoccupation with the financial aspects of development projects – enter the utility function of the manager of a public enterprise.

Though it was not designed to answer this particular question,[20] the public choice approach to bureaucracy does nevertheless throw considerable light on it. For what this approach in fact postulates are a variety of mechanisms through which bureaucrats benefit – in terms of 'salary, perquisites of the office, public reputation, power [and] patronage'[21] – from *larger budgets*. Each of these mechanisms may be said to provide a possible motivational basis for the bureaucratic choice of techniques since the preoccupation with securing foreign exchange in that approach can be directly related to the more general objective of budget maximization. And with these specific mechanisms in mind, it becomes much more meaningful to examine the literature referred to above, that deals with the political role of the bureaucracy in post-colonial Africa.

This same literature also suggests however that it would be a mistake to focus exclusively on the rather narrow goals pursued by the bureaucracy, however predominant this class may have become in overall political economy terms. For in Africa, as elsewhere in the Third World, one sometimes needs to make the distinction:

> between the top political leadership representing the state (let us call it the state elite) which takes the general political decisions and the hierarchy of agents, the bureaucracy, which is supposed to implement those decisions. The process of implementation often generates various kinds

> of rental income, which, to a significant extent, accrues to the bureaucracy and the latter forms a pressure group to secure this income flow, with goals that are often narrower than those set by the state elite. The impulses that shape major policies and actions by the latter are fuelled not merely by motives of self-aggrandizement but quite often also by . . . its 'conception of the national interest'. . . . In many cases of state directed industrialization this leadership genuinely considers itself, as the trustee of the nation's deeply held collective aspirations . . . one, though not the only, form these aspirations often take is to strive for rapid economic development.[22]

The importance of this distinction does not go entirely unnoticed in the discussion (in Part I) over the bureaucratic choice of techniques. However, whereas that discussion is framed largely in terms of the *organizational* problems of the relationship between agents and their principals, Chapters 7 and 8 emphasize the *political* aspects of that relationship. In Chapter 3, for example, the divergence between what was intended by the 'top elite' and what actually occurred is ascribed, somewhat abstractly, to difficulties in the formulation of a national objective function with which to hold the bureaucracy to account. In Chapter 7, in contrast, the same discrepancy between intentions and outcomes is viewed more concretely as the result of an intense and protracted political struggle between the 'top elite' on the one hand and the bureaucracy on the other. In Tanzania, as in certain other African countries, this conflict was marked by *active* efforts on the part of the latter to insulate itself from the controls (over public enterprises) that the top elite and other regulatory agencies of the state sought to exercise.

One particular consequence of these bureaucratic strategies has been a marked lack of success in bringing about a more *decentralized* bureaucratic state in Africa. And as argued in Chapter 8, this unyielding feature of the bureaucracy has had and will continue to have, a powerful influence on the pattern of technology and industrialization in African countries. For, among other reasons, a highly centralized bureaucracy is most unlikely to display the degree of attention to and interest in the types of small-scale technologies that an alternative and more egalitarian development strategy would entail.

None of the foregoing is meant to suggest, however, that the narrow interests of the bureaucratic class alone are responsible for the observed technological biases in the public sector. On the contrary, the pursuit of 'rapid economic development' has *itself* exerted an influence in the same direction. This influence, it should be emphasized, is not inherent in the goal itself, as the very different technological experience in the East Asian newly industrializing countries amply demonstrates. Rather, it has to do with the specific manner in which African leaders set about the pursuit of this objective.

What is probably most compelling in this respect is Hyden's contention that 'African leaders have wanted to make historical shortcuts'.[23] They have tried to replace conventional development sequences with rapid structural changes that were explicitly designed to break with the past. In this accelerated vision of growth, with its emphasis on historical discontinuities (a process that Hyden aptly refers to as 'running while others walk'), large-scale production based on imported technology and other inputs came to acquire a particular appeal. For, to a greater extent than was thought possible on a smaller scale, this form of production held out the hope of achieving the appreciable and very rapid increases in output that were desired (in some cases, this general view was closely related to and reinforced by particular elements of socialist ideology, which tended to favour advanced techniques in order 'to facilitate surplus extraction').[24]

In this calculation, other considerations, it seems, exerted little or no influence. Questions of efficiency, for example, could hardly be brought to bear on enterprises that were so heavily favoured by the prevailing strategy of import substitution, or, more accurately, of import 'reproduction'.[25] Nor, apparently, was much attention paid to the question of whether the favoured enterprises were actually contributing to the acquisition of indigenous technological capabilities (Mytelka, for example, refers to the Côte d'Ivoire experience with industrialization as a 'process in which opportunities for technological mastery are systematically missed'[26] and similar observations about other countries could easily be cited). In their neglect of these and other factors, African leaders were thus effectively allowing short-run policy goals to 'justify means' and in so doing they were (albeit in part perhaps unwittingly) making long-run economic growth all the more unlikely. For not only did the emphasis on the growth of large-scale enterprises fail to produce corresponding gains in terms of technological capabilities, but it also led to the neglect of the considerable potential that the small-scale sector had to offer; a neglect, one should stress, that was no less pronounced in agriculture, where 'crash modernization', with its emphasis on tractors and large farms reduced 'the possibility of successfully pursuing a strategy aimed at widespread modernization of smallholder farming based on fuller and more efficient utilization of their "internal inputs" of labor and land and by a sequence of innovations which involves gradually expanded use of purchased inputs which complement the relatively abundant internal, farm-supplied resources'.[27]

RECENT POLICY REFORMS

Beginning in the mid-1980s, a series of important, and by now familiar, policy reforms were introduced in most countries of Sub-Saharan Africa by the World Bank and the IMF. Particularly since much of our case-study material pertains to the period preceding those reforms, there is an obvious possibility that the problems we have identified in this book are no longer relevant. If, for example, the privatization component of structural adjustment had been as successful as some policymakers had hoped, the public sector itself would be considerably less prominent than was mostly the case during the 1970s and early 1980s. Or, even with no significant change in the degree of public ownership in the industrial sector, the various policy reforms that were pursued in the name of economic liberalization might – directly or indirectly – have redressed many of the technological problems described earlier in this introduction, according to at least some interpretations of the causes of those problems.

Yet, for all the influence that these and other recent policy reforms were supposed to exercise over the extent and behaviour of the African state, there is in fact very little evidence to suggest any appreciable obsolescence of the technological and other concerns that earlier experience in the public sector had raised so prominently. There are certainly very few African countries in which privatization has not encountered formidable obstacles of one kind or another (as indeed one would have predicted on the basis of even a cursory political economy analysis of the period preceding the introduction of structural adjustment policies). As a consequence, even the most recent data suggest that in many such countries, ownership patterns have not been altered to any appreciable extent. A World Bank Study of 29 African countries, for example, comes to the conclusion that:

> Only a handful of countries have divested more than 40 per cent of their enterprises. And half the countries have been extremely slow to privatize any enterprises. In Kenya there have been almost no sales in ten years. A few countries have even expanded their public enterprise sectors. In Burundi the public enterprise sector grew during the adjustment period, with five firms divested but twelve new ones created. In Côte d'Ivoire the number of public enterprises rose from 113 in 1977 to 140 in 1990, despite the privatization of some 30 enterprises in the mid-1980s.[28]

Nor it seems has the economic and technological behaviour of the large number of remaining public enterprises necessarily been altered by even the successful introduction of policies to liberalize domestic and international markets. Here again the recent World Bank study is worth citing and in particular its overall assessment that:

> the efforts at reform have been disappointing . . . Few 'rehabilitated' public enterprises have made significant and lasting improvements, and many are being rehabilitated for a second and third time. Another worrisome trend is the ongoing, large annual investment in these public enterprises (often financed in part by donor funds), even though they are generating low or negative returns.[29]

For example, in at least one African country that features prominently in this book, Tanzania, the technological and economic behaviour of enterprises in the public sector appears to have remained largely unchanged in recent years, in spite of the vigour with which many structural adjustment policies have been pursued there.[30] To this extent, therefore, the policy issues that are identified in most of the following chapters are no less relevant, or less urgent, than they were in the 1970s and the early 1980s.

Notes

1. See Swanson and Wolde-Semait (1989) and Nellis (1986). According to the latter, 'state-owned enterprises dominate the economies of many African countries. . . . PEs [public enterprises] account for over 90 per cent of manufacturing value-added in Ethiopia, almost 80 per cent in Somalia, over 50 per cent in Zambia and 40 per cent in Cameroon' (pp. 7–8). Though some of these totals may have been somewhat reduced in recent years as a result of privatization schemes, in most African countries the public sector remains extremely important (see note 28 below).
2. This particular methodological proclivity is not at all uncommon in microeconomics; on the contrary, it forms part of the very notion of rationality that has grown up and spread so pervasively within the economics discipline. Simon (1979) refers to this as 'substantive rationality' and he contrasts it with the 'procedural rationality' that has developed within psychology. Behaviour is described as substantively rational 'when it is appropriate to the achievement of given goals within the limits imposed by given conditions and constraints' (1979, p. 67). On the other hand 'Behaviour is procedurally rational when it is the outcome of appropriate deliberation. Its procedural rationality depends upon the *process* that generated it' (1979, p. 68, emphasis added). It follows from these contrasting concepts of rationality that the latter has little or nothing to offer the former. What we advocate below is in effect a move away from substantive towards

procedural rationality: towards a concern, that is to say, with the detailed and complex procedures that *underlie* observed outcomes.
3. Williams (1976).
4. Kaplinsky (1990), p. 93.
5. Johnston (1981), p. 21.
6. As shown in the following chapter.
7. World Bank (1982), p. 5.
8. Ibid., pp. 10–31.
9. Ibid., p. 5.
10. Hirschman (1984), p. 91.
11. Ibid.
12. Williams (1975).
13. See for example Wells (1975), Winston (1979) and Hawrylyshyn (1978).
14. Williams (1975), p. 8.
15. Ibid., (emphasis added).
16. Chazan *et al.* (1988), p. 51, emphasis added.
17. Ibid., p. 53.
18. Bates (1981), p. 53.
19. Gallagher (1991).
20. On the evolution of this approach see Dunleavy (1991) and Mueller (1989).
21. Niskanen (1971), p. 38.
22. Bardhan (1988), p. 65.
23. Hyden (1983), p. 123.
24. Green (1978) shows that this ideology had some influence over the choice of an automated, large-scale bakery in Tanzania.
25. Mytelka (1992).
26. Ibid., p. 260.
27. Johnston (1981), p. 36.
28. World Bank (1994), p. 104. The limited extent of privatization in Africa is also discussed in several other recent publications, such as Stewart, Lall and Wangwe (1992); Sandbrook (1993); *The Economist*, 1 May 1993; and Due (1993).
29. World Bank (1994), p. 109.
30. See Chapter 7 below for a discussion of this case.

References

P. Bardhan (1988). 'Alternative Approaches to Development Economics', in H. Chenery and T.N. Srinivasan (eds), *Handbook of Development Economics*, North-Holland.

R. Bates (1981). *Markets and States in Tropical Africa*, University of California Press.

N. Chazan, R. Mortimer, J. Ravenhill and D. Rothchild (eds) (1988). *Politics and Society in Contemporary Africa*, Lynne Rienner.

J. Due (1993). 'Liberalization and Privatization in Tanzania and Zambia', *World Development*, vol. 21(12).

J. Dunleavy (1991). *Democracy, Bureaucracy and Public Choice*, Harvester Wheatsheaf.

M. Gallagher (1991). *Rent Seeking and Economic Growth in Africa*, Westview.

R. Green (1978). 'The Automated Bakery: A Study of Decision Taking Goals, Processes and Problems in Tanzania', Institute of Development Studies, Sussex University, Working Paper no. 141.

O. Hawrylyshyn (1978). 'Capital-Intensity Biases in Developing Country Technology Choice', *Journal of Development Economics*, vol. 5.

A. Hirschman (1984). 'A Dissenter's Confession: "The Strategy of Economic Development" Revisited', in G. Meier and D. Seers (eds), *Pioneers in Development*, Oxford University Press.

G. Hyden (1983). *No Shortcuts to Progress*, Heinemann.

B. Johnston (1981). 'Farm Equipment Innovations and Rural Industrialisation in Eastern Africa: An Overview', ILO, World Employment Programme, Working Paper no. 80.

R. Kaplinsky (1990). *The Economies of Small*, IT Publications.

D. Mueller (1989). *Public Choice II: A Revised Edition*, Cambridge University Press.

L. Mytelka (1992). 'Ivorian Industry at the Cross-Roads', in F. Stewart, S. Lall and S. Wangwe (eds), *Alternative Development Strategies in SubSaharan Africa*, Macmillan.

J. Nellis (1986). 'Public Enterprises in Sub-Saharan Africa', World Bank Discussion Papers.

W. Niskanen (1971). *Bureaucracy and Representative Government*, Aldine-Atherton.

R. Sandbrook (1993). *The Politics of Africa's Economic Recovery*, Cambridge University Press.

H. Simon (1979). 'From Substantive to Procedural Rationality', in F. Hahn and M. Hollis (eds), *Philosophy and Economic Theory*, Oxford University Press.

D. Swanson and T. Wolde-Semait (1989). 'Africa's Public Enterprise Sector and Evidence of Reforms', World Bank Technical Paper no. 95.

F. Stewart, S. Lall and S. Wangwe (eds) (1992). *Alternative Development Strategies in SubSaharan Africa*, Macmillan.

L. Wells (1975). 'Economic Man and Engineering Man: Choice of Technology in a Low-Wage Country', in C.P. Timmer et al., *The Choice of Technology in Developing Countries*, Harvard Studies in International Affairs.

D. Williams (1975). 'National Planning and the Choice of Technology: the case of Textiles in Tanzania', Economic Research Bureau Paper no. 75.12, University of Dar-es-Salaam.

G. Winston (1979). 'The Appeal of Inappropriate Technologies: Self-Inflicted Wages, Ethnic Pride and Corruption', *World Development*, vol. 7(8/9).

World Bank (1982). *Technology and Science in World Bank Operations*, Washington, DC.

World Bank (1994). *Adjustment in Africa: Reforms, Results and the Road Ahead*, A World Bank Policy Research Report, Oxford University Press.

Part I
The Bureaucratic Choice of Techniques

1 The Growth and Technological Performance of Public Sector Institutions

This chapter is closely linked to the previous one because it documents in some detail the systematic tendencies in the technological behaviour of public enterprises that were mentioned there. It is also closely related to the later chapters because of the emphasis that it places on external capital in financing the extraordinarily rapid growth of the public sector. Indeed it is with the nature of, and the interactions between, these two distinguishing features of industrialization in Africa that the subsequent chapters of this book are mainly concerned.

THE GROWTH OF PUBLIC ENTERPRISES

As one might expect, countries which after independence were strongly committed to socialism (and in particular to what is sometimes referred to as 'African socialism') witnessed a sharp increase in the share of state ownership in the industrial sector. In Tanzania, for example, the share of parastatal enterprises in manufacturing value added rose from 5 to 31 per cent, over the period from the mid-1960s to the late 1970s (see Table 1.1). The more disaggregated data shown in Table 1.2 illustrate just how rapid was the growth in certain key sectors during the last few years of this period. In Ghana, over a similar period, the share of enterprises with some degree of public ownership rose from 21 to 53 per cent (see Table 1.1).

The expansion of the public sector in industry was not however confined to countries in which state ownership had such powerful ideological overtones. 'For example, even though Ivory Coast and Nigeria were relatively open to foreign investment, the share of their public sectors in industrial equity capital rose from 10 and 22 per cent in the 1960s to 24 and 38 per cent, respectively, in 1975'[1] (see Table 1.1). In these and other cases where the prevailing ideology was less obviously predisposed to the expansion of public enterprises, different arguments came into play to exert pressure in the same direction. Two of these, according to Hyden, are of particular importance in explaining the widespread expansion of parastatals that took place in Africa during the 1960s and 1970s: namely,

Table 1.1 Public enterprise expansion in the manufacturing sector: PE share of manufacturing value added (per cent)

	1960s		*1970s*	
	Beginning	*End*	*Beginning*	*End*
Ghana				
State-owned public enterprises	13.3	16.0	17.2	27.1
Joint state/private	7.9	19.7	18.0	25.6
Sub-total	21.2	35.7	35.2	52.7
Private	78.9	64.3	64.8	47.3
Zambia				
Public/parastatal	n.a.	n.a.	53.2	56.4
Private	n.a.	n.a.	46.8	43.6
Tanzania				
Parastatal	5.0	22.5	32.2	31.0
Private	95.0	77.5	66.8	69.0
Kenya				
Public enterprise	15.5	13.2	18.0	n.a.
Private	84.5	86.8	82.0	n.a.

	PE share of industrial equity capital	
	Early 1960s	*Mid-1970s*
Côte d'Ivoire	10.0	24.0
Nigeria	22.0	38.0

Source: Steel and Evans (1984), p. 62; Swanson and Wolde-Semait (1989).

Table 1.2 The growth in capacity of selected manufacturing industries in Tanzania

*Industry**	*Units*	*Capacity 1976*	*Capacity 1980*
Textiles	Sq. m. (million)	90	200
Cement	000 tons	340	1100
Sugar	000 tons	115	195
Shoes	million pairs	6	14

* These industries are dominated rather than owned entirely by the government.

Source: Wangwe (1983) cited in Mukandala (1988), p. 129.

the control by foreigners of these economies and the comparative weakness of domestic capitalism. More specifically, he suggests that:

> In the absence of an entrenched class of local capitalists, many governments felt a need to resort to public enterprises as the only alternative to counteract the influence of foreign capital and to accelerate the development of local resources. The public enterprises would take the role played by private entrepreneurs in other countries and thus help to harness, mobilize and exploit resources which would otherwise lie idle or be developed only by foreigners.[2]

Yet, for all the importance thus apparently attached to the question, in neither the socialist nor the other types of development strategy was the growth of public enterprises accompanied by a diminished reliance on foreign capital. On the contrary, state ownership itself served on occasion to attract certain forms of (private) foreign capital.

In the Tanzanian context, for example, Clark argues that 'Foreign capital was attracted when the government initiated projects. International firms, anxious to sell equipment, were willing to have joint ventures, and provide suppliers' credit to areas of the economy where they would have never gone without the security of government involvement and of government backed loans.'[3] Indeed, during the First Five Year Plan, parastatals were able to raise almost the same amount in private foreign loans as the ministries had received in foreign aid during the same period.[4] Similarly, during the sixties in Ghana – a period that was marked, as we showed in Table 1.1, by a rapid growth in state ownership – the level of foreign indebtedness increased by a factor of 20–25, mostly on account of what Grayson refers to as a 'phenomenal' use of suppliers' credits.[5] These were used, among other purposes, to finance all public enterprises in the manufacturing sector. Grayson observes that the suppliers' credits 'favored public enterprises. In fact, virtually all such credits to English-speaking West Africa were extended to public or quasi-public entities; and all such credits in Ghana went to the public sector'.[6] Still another example that can be cited is Côte d'Ivoire, where state participation in the industrial sector rose from 18.2 per cent of social capital in 1971 to 30.4 per cent in 1979 and 52.2 per cent in 1981. Yet, according to Mytelka, foreign loans and suppliers' credits 'accelerated' over this period. 'For the foreign investor, in the context of increased state equity participation coupled with loans guaranteed by the state, the resort to foreign borrowing constituted a key element in a strategy of risk reduction.'[7] In the Ivorian sugar industry, for

Table 1.3 Net disbursements of ODA from bilateral and multilateral sources to Sub-Saharan Africa 1973–83 (in $ million at 1984 prices and exchange rates)

	1973	*1983*	*1973–83 Average annual growth (per cent)*
Sudan	70	937	124
Ethiopia	123	332	17
Tanzania	185	581	21
Kenya	177	392	12
Somalia	89	320	26
Mali	132	210	6
Senegal	146	315	12
Zaire	258	308	2
Zambia	84	312	37
Mozambique	–	206	
Niger	131	171	10
Ghana	76	108	4
Burkina Faso	105	180	7
Mauritania	57	172	20
Uganda	41	134	23
Rwanda	72	146	10
Madagascar	97	175	8
Zimbabwe	2	204	1010
Cameroon	113	126	1

Source: DAC, *Development Co-operation*, OECD, Paris, 1986, p. 183.

example, where the state was a major investor, foreign loans amounted to more than 60 per cent of the capital invested.[8]

Foreign aid donors too exhibited a keen interest in the growing numbers of public enterprises. Indeed, Table 1.3 shows that the decade beginning in 1973 witnessed rates of growth in aid disbursements of more than 10 per cent in numerous African countries; much of this aid it seems went to finance investment in the public sector.[9] Consider for example the Zambian case. By the end of the 1970s, public investment accounted for almost the entire amount invested in the formal sector of that country.[10] And, as Table 1.4 indicates, no less than 65 per cent of that total amount was financed (in 1984) by foreign aid. Table 1.4 also indicates that in several of the poorest countries, aid financed not only all investment expenditure but part of consumption as well.

The heavy involvement of foreign capital with public enterprises in Africa, that has just been described, did not exclude the World Bank. On the contrary, the Bank has been a major investor even in countries that

Table 1.4 Foreign aid and domestic investment, 1984: IDA eligible countries

	Net official development assistance (ODA): disbursements		
IDA eligible country	*Per capita (dollars)*	*As percentage of gross domestic investment*	*Bilateral (% of total)*
Benin	19.8	125.2	51.0
Senegal	52.0	91.6	73.7
Malawi	23.3	80.3	32.6
Guinea-Bissau	61.2	134.4	55.2
Sierra Leone	16.4	68.6	37.0
Sudan	28.7	62.6	50.1
Central African Republic	45.5	161.0	59.9
Kenya	21.9	33.4	68.3
Ghana	16.1	45.8	44.1
Mauritania	99.1	103.8	40.7
Zambia	36.6	64.4	75.8
Tanzania	26.0	50.0*	72.9

* Refers to foreign finance as a proportion of government expenditure for 1980–5.
Source: World Bank, *Financing Adjustment with Growth in Sub-Saharan Africa, 1986–90*, 1986, Table 18. For note (1) S. Wangwe (1992).

were widely regarded as pursuing a socialist strategy of development. During the 1970s in Tanzania, for example, the Bank invested some 140 million US dollars in four very large public enterprises in the industrial sector, including the ill-fated shoe factory in Morogoro.[11] In addition, substantial amounts were lent to the public sector via the loans that were made to development finance corporations (DFCs) in that country.[12]

Although it may seem to be, this degree of involvement with the public sector in Tanzania and elsewhere in Africa was not entirely out of keeping with either the Bank's philosophy or its overall operations in the 1960s and 1970s. For during this period not only did the Bank invest in public sector industrial projects of an import-substituting kind in a wide range of developing countries, but it also made loans to state-owned DFCs in such magnitude that by 1978 these corporations accounted for more than 10 per cent of the combined loans granted (by the IBRD and IDA).[13] Indeed, on the basis of these and other observations, one commentator has gone so far as to argue that 'whatever the original profession of intent or the initial pattern of practice, the World Bank has been an underwriter of state-led growth in developing countries during much of its existence. The focus on state-owned enterprises was but one means of facilitating an expanded role for the state in a variety of areas.'[14]

During the 1980s, of course, the World Bank's stance towards the public sector in Africa (and elsewhere) underwent an important change and it was accompanied not only by a new set of policies towards this sector but also by an altered view of the role of states and markets in development thinking itself.[15] We shall return to these issues in a later chapter. For the moment, however, we need only reiterate the point made in the previous chapter, that in spite of recent structural adjustment and other policy reforms in Sub-Saharan Africa, state-owned enterprises continue to dominate the industrial sector of most such countries.

THE TECHNOLOGICAL PERFORMANCE OF PUBLIC ENTERPRISES

Public enterprises in the industrial sector of most African countries have generally failed to contribute towards either efficiency or other major economic goals, such as employment creation. This conclusion is supported by a large body of empirical evidence, covering a range of countries and industrial branches. What is somewhat less clear, though, is the extent to which the *technological* behaviour of these enterprises (as opposed to other factors) has contributed to this disquieting experience. Our attempt to answer this question comes essentially in two parts. The first seeks to discern the significance of the technological and related variables that bear on overall economic efficiency. The second part then describes the technological behaviour of public enterprises in terms of these same variables. By combining these two pieces of evidence one can thus infer the extent to which technological behaviour in the public sector may have contributed to the typically inefficient pattern of industrialization in Sub-Saharan Africa.

The Choice of Techniques

'The degree of inefficiency in African industry', writes Lall, 'seems to rise with the degree of capital . . . intensity of the facilities set up.'[16] Surveys of public (and other) enterprises in three different African countries tend to confirm this view. In his study of Ghana, for example, Steel concluded that capital intensity was a statistically significant explanation of the high domestic resource cost (DRC) which he observed.[17] In Zambia state-owned enterprises with 'relatively high DRCs (above 2.0) tend to have higher capital intensity . . . a one-per cent increase in the capital/labor ratio is associated with an increase of 0.2 per cent in the DRC'.[18] Or again, in Tanzania, 'Efficiency of manufacturing is also negatively affected by higher capital intensity'.[19]

Independent variables that are distinct from, but related to, factor-intensity often also play an important explanatory role. Firm size is one such variable that is statistically significant in explaining the high DRC in Steel's study of Ghana and that also features prominently in the Tanzania case, where 'firms employing between five and 100 workers had a domestic resource cost of 1.07; while for firms employing between 100–250 and 250–1500 domestic resource costs were 5.54 and 20.54 respectively'.[20] Import intensity is another variable that often exhibits a systematic relationship not only to factor-intensity, but also to the overall efficiency of a firm, as this is measured by DRC. In particular, methods of production that are relatively import-intensive often tend also to be associated with capital rather than labour-intensive technologies, and to increase rather than decrease domestic resource cost. In Zambia, for example, 'As in other developing countries, high-cost production is associated with the characteristics of high capital and import intensity. . . . Both of these variables are significant in explaining variations in the DRC between firms.'[21]

Table 1.5, which is also based on survey evidence, describes the technological behaviour of public enterprises in these same countries (with the addition of evidence from Kenya) and in terms of the same variables: factor intensity, firm size and import intensity. Clearly, there is a systematic tendency for public enterprises in these countries to exhibit precisely the type of technological behaviour that is known to produce economic inefficiency – namely, high capital and import-intensity and (relatedly) large-scale production. There is also some evidence to suggest that public enterprises in Kenya, Ghana and Tanzania may be more capital-intensive than privately owned firms in those countries. This evidence is shown in Table 1.6, but it should be considered as suggestive rather than conclusive, one reason being that the observed differences in behaviour are not statistically significant. Furthermore, at least one of the country studies (Tanzania) can be criticized on the grounds that firm size was not adequately controlled, so that the higher capital intensity associated with the public enterprise may be a question of size (e.g. diseconomies of scale) rather than of ownership.[22]

What is also not entirely clear from the associations described in the previous paragraph is how the technological, or technologically related variables – of scale and factor-intensity – are actually translated into efficiency terms (that is, into DRC measures). For there are in principle a

Table 1.5 Public enterprise and the choice of technology, selected African countries

Country	*Description*
Zambia	'A major area of concern for the public/parastatal sector remains the use of appropriate investment criteria for selection of new projects. Past public sector investments have been characterized by high (and frequently excessive) capital intensity and dependence on imports' (World Bank, 1984, p. v).
Kenya	'The parastatal client firms tend to be large, capital intensive import intensive and almost exclusively oriented toward a protected, over priced local market' (Hopcraft and Oguttu, 1982, p. 193).
Ghana	'The strong preference for modern, capital-intensive technology by the state, coupled with the lower cost of capital it faced, made it possible for the state owned firms to be highly capital intensive compared with the rest. . . . these results raise the disturbing possibility that the involvement of the state in direct industrial production activities solely or jointly with foreign firms . . . may be less helpful to the developing countries than had been considered' (Ahiakpor, 1986, p. 159). 'Evidence on individual projects supports other data in indicating the public sector to be particularly prone to capital intensity' (Killick, 1978, p. 228).
Tanzania	'A study of over 300 industrial units in ten major industries found that labour-intensive, low unit costs and frequently small-scale techniques are being employed in Tanzania' (Perkins, 1983, p. 218). Yet, 'Despite the rhetoric, Tanzania's industrialization programme has, in general, promoted the establishment of [publicly owned] enterprises using large-scale capital-intensive, often technically, and almost invariably economically inefficient techniques' (ibid., p. 231). 'There is a strong tendency for parastatals to adopt the most advanced techniques, and build the most modern buildings. In manufacturing this tendency is exhibited as an inclination towards large-scale, capital-intensive, "transformation" projects' (Clark, 1978, p. 125).

Note: As it was originally formulated, this table involved only the comparison between the Kenya and Tanzania cases that appears in Table 3.1. below

Table 1.6 Technology choice in public vs. private enterprises, selected African countries

Country	*Result of comparison*
Ghana	'Though they are not statistically significant, the other results from the pooled sample give the impression that wholly state owned firms are more capital intensive than private foreign firms (2.1 to 1) while the mixed private and private local firms are less capital intensive than the private foreign firms.'
Kenya	In comparisons between firms in the same industry producing an identical or similar product 'on the whole parastatal firms are more capital intensive than other firms'.
Tanzania	In ten major industries, it was found that 'In every sector except printing or publishing, Government or joint-owned firms operated more capital intensive and less capital productive techniques than private firms.'

Sources: For Ghana, Ahiakpor (1986, p. 155); for Kenya, Hopcraft and Oguttu (1982, p. 191) and for Tanzania, Perkins (1980, p. 197).

number of different ways in which these influences can be exerted. Relatively high DRCs, for example, can be the result of poor technological choices (that is, allocative inefficiency) or of inefficient use of *given* technologies (that is, X-inefficiency) or both of these.[23]

That said, however, there is certainly some evidence that inefficient choices have occurred over a wide range of public enterprise activity, where alternative technologies existed (as reported for example by Perkins,[24] Bagachwa[25] and others[26] for the Tanzania case). There is also a compelling body of evidence gleaned from a variety of case-studies in the choice of technology literature, that factor substitution possibilities do exist across a wide spectrum of manufacturing industries (such as food processing, textiles and cement) and Pack has shown that when these possibilities are aggregated, the macro-economic effects of alternative technologies on employment and output are very substantial.[27]

More specifically, using market prices that are thought to be typical of most Sub-Saharan African countries and on the basis of various other assumptions, Pack compares, for each of nine manufacturing industries, two extreme technical options, namely, the most capital-intensive and the 'appropriate technique'. The former he suggests 'may be thought of as a turnkey plant designed by a consulting engineering firm or, a diversified equipment producer'.[28] The latter, on the other hand, is defined as

Table 1.7 Employment and output implications of alternative techniques at the macro-economic level

Technique	*Value added ($ million per annum)*	*Employment (all workers)*	*Capital–labour ratio ($ per worker)*
Appropriate	624	238,678	3,771
Capital-intensive	364	58,017	15,513

Source: Pack (1982), Table 3.

the technique which maximizes the present discounted value per unit of capital cost. Pack then proceeds to estimate for each such industry the differences in value added, employment and capital per worker that would result from investing a given amount ($100 million) in the two technical options that have just been described. When they are aggregated over all the industries in his sample, these calculations yield the results shown in Table 1.7.

On average, therefore, the 'appropriate' technique generates not only a substantially greater amount of employment than the capital-intensive alternative, but more value added as well. These differences cannot of course be taken to represent in any precise way the aggregate output and employment losses from the technical choices now being made by public enterprises in the context of Sub-Saharan Africa. For one thing, these enterprises are not always to be found in the sectors to which the calculations refer. And even where they do exist, public enterprises may not always choose the capital-intensive technique in the particular sense defined in the previous paragraph. Yet, it does appear that public ownership still pervades much of African industry and that technical choices made by state-owned enterprises tend towards the capital-intensive rather than the 'appropriate' end of the spectrum of available factor intensities (and as is also suggested below, towards a packaged, rather than an unpackaged form of technology transfer). To this extent, therefore, it seems reasonable to conclude that the estimates contained in Table 1.7 at least approximate the macro-economic consequences of the technological behaviour exhibited by public enterprises in the industrial sector. That is, the estimates in the table approximate the difference between what actually occurs and what might conceivably take place, were firms in the public sector to shift from the one form of technology to the other over a wide range of manufacturing industries.

This conclusion, moreover, is based only on direct (or partial equilibrium) effects. That is, it is based on aggregating the partial-equilibirum effects in a number of different sectors. If the indirect (or economy-wide) effects were also to be included, it is quite possible that the output and employment differentials estimated by Pack would become even more pronounced. Khan and Thorbecke's study of six disaggregated industrial sectors in Indonesia, for example, found that the direct and indirect linkages generated by relatively labour-intensive techniques were in most cases appreciably greater than those generated by the more modern, capital and import-intensive alternatives.[29]

The Acquisition of Technological Capabilities in the Public Sector

Important though it may be, the choice of technology is of course not the only aspect of technological behaviour that needs to be considered. On the contrary:

> The ability to make independent technological choices, to adapt and improve upon chosen techniques and products, and eventually to generate new technology endogenously are essential aspects of the process of development. The process may be described as the accumulation of technological capacity; it is at least as important to economic development as the accumulation of capital.[30]

In Africa, as elsewhere in the developing world, the efficacy of this process is heavily influenced by the form in which technology is transferred to the industrial sector. It is also influenced by the extent to which recipient enterprises and other institutions in the public sector actively seek to assimilate the technology that is acquired, in very large measure, from the developed countries. From both these points of view, one can again (as above with the choice of technology) discern certain rather systematic patterns of technological behaviour in the public sector.

With respect, first, to the mode of technology transfer, the available case-study evidence suggests that this has occurred typically in a highly packaged form. The brewing industry in Nigeria, for example, is predominantly state owned: in nine of the thirteen firms in the industry, the state owns the majority of shares. All of these enterprises were built on a 'turnkey' basis: that is, by means of a specially packaged form of technology transfer which requires that the foreign collaborator perform all the tasks that are needed for production to begin.[31] Similarly, in a study of industrial projects in Tanzania, Wangwe found that most turnkey projects occurred in the public sector. In particular, 'While the public sector enterprises constitute 55.2 per cent of the sample, they constitute 77.8 per cent of all packaged turnkey projects. It appears that *it is more in the public sector that packaged turnkey projects have been adopted*.'[32] Still other examples of packaged forms of technology transfer to public enterprises (albeit not necessarily in the specific turnkey form), are provided by Mytelka (for Côte d'Ivoire),[33] Grayson and Adei (for Ghana),[34] Juma (for Kenya),[35] Barker et al. (for Tanzania)[36] and Ohiorhenuan and Poloamina (for Nigeria).[37]

Most of the case-studies that have just been cited are also at pains to emphasize just how little attention was paid by state institutions to the acquisition of technological capabilities during the process in which technology was transferred. State intervention in the Ivory Coast during the 1970s, for example, appears to have been 'primarily designed to reduce the risks to foreign investors and not to influence the choice of technology or its subsequent assimilation'.[38] A similar attitude was discovered by Barker *et al.* in Tanzania, where the National Development Corporation 'was quite happy to accept turnkey projects offered by the foreign partners . . . instead of concentrating on building up [indigenous] technological expertise among nationals, which is a time consuming and difficult business'.[39] In Ghana, a detailed case-study of the state-owned Bonsa Tyre Company – which involved the transfer of technology from the Firestone Corporation – indicated that 'technological considerations were at best secondary to the government'.[40] Even in Nigeria, which is sometimes described as being relatively attentive to the acquisition of technological capabilities, a recent study concludes that 'government policies merely facilitate production rather than encouraging utilisation of local resources',[41] while a separate analysis of the brewing industry in that country comes to much the same conclusion.[42]

With encouragement, therefore, from neither the typically packaged form of technology transfer, nor the state itself, it is hardly surprising that the acquisition of technological capabilities by public enterprises has been as limited as is suggested by Table 1.8 (which contains the main conclusions reached by the authors of the various case-studies referred to above).

Part of what seems from Table 1.8 to have occurred in the public sector can be described (and criticized) in terms of Hirschman's concepts of 'trait-taking' and 'trait-making'.[43] The former refers to the practice of taking as given certain traits (e.g. corruption, values, capabilities) that are relevant to the outcome of a project but which, for one reason or another, are difficult to alter. 'Trait-making' on the other hand, sees the need to alter traits as part of the project itself, for, 'the project, *if it is successful*,

Table 1.8 Public enterprise in industry and the acquisition of indigenous technological capabilities, selected African countries

Country	*Sector(s)*	*Observations*
Côte d'Ivoire	Textiles	'the state . . . took shares in six of the eight textile firms. Yet in the context of continued reliance on foreign capital and technology suppliers, this new role for the state had little impact on the building of indigenous technological capabilities through the technology transfer process.'
	Sugar	'The history of these sugar complexes is replete with design errors and missed opportunities for learning and technological mastery.' Five of the remaining complexes that were built after 1972 'could have served as a learning base . . . But these were built in rapid succession allowing no time for Ivorians to move up the learning curve . . . The possibilities for cumulative learning in the bargaining process were further reduced by the choice of six different technology suppliers/managers and the holding of negotiations nearly simultaneously with five of these.'
Nigeria	Beer brewing	'The greatest disadvantage of the turn-key arrangement . . . is that is offers little opportunity for indigenous personnel to participate in the various stages of implementing the project. It thus denies this personnel the opportunity to learn by doing . . . This has been the case in the Nigerian brewing industry.'
	Cement	'Despite the fact that the public sector is the major owner of all the cement plants, non-cooperation among governments has made it impossible for the problem factories to look to other successful plants for technical assistance. Consequently, the country lost an opportunity for successful acquisition of cement technology.'

continued

Table 1.8 (continued)

Country	*Sector(s)*	*Observations*
Ghana	Tyres (Bonsa Tyre Company)	'On the whole, performance . . . deteriorated after 1983 when the government became the sole owner of the firm and management was transformed to nationals. Weakness in technological capacity in the firm has been identified as a major factor accounting for the poor performance.' More generally in Ghana, 'where the government went into partnership with TNCS, no attempt was made to use past ownership to secure the building of indigenous capacity. The country had forfeited the opportunity to acquire the necessary know-how to run the companies.'
Tanzania	Textiles, metalworking, non-metallic and leather	'In the process of implementing turnkey projects [in the public sector] the local manpower is "freed" from participating in those preparatory and investment decision activities which have considerable learning effects . . . This way the opportunity to increase the supply of managerial and technical skills through learning-by-doing is foregone.'

Sources: Côte d'Ivoire, Mytelka (1985, p. 87) and (1992, pp. 251 and 252 respectively); Nigeria, Akpakpan (1986, p. 671) and Ohiorhenuan and Poloamina (1992, p. 305) respectively; Ghana, Adei (1990, pp. 1511 and 1504); Tanzania, Wangwe (1986, p. 15).

will be valuable not only because of its physical output but even more because of the social and human changes it will have wrought'.[44]

The choice between these two alternatives is described by Hirschman as nothing less than the 'basic dilemma of project design in an underdeveloped country'. On the one hand, if the project accepts certain undesired traits, it runs the risk of missing opportunities to change these and it may even reinforce them. On the other hand, if the project seeks to alter what are regarded as undesirable traits, it also runs a risk, namely, that if the alterations do not in fact occur (or at least do not occur rapidly enough),

Table 1.9 Trends in the formation of science and technology policy-making bodies in Africa 1973–86

	1973	*1979 Increase*	*Total*	*After 1979*	*Total 1986*
Ministry of Science or Ministerial Science Policy Council	5	+4	9	+18*	27
Science planning body in general	12	+6	18	+2	20
Multisector body for coordinating scientific research	18	+6	24	+4	28
Natural science research	2	+14*	16	+9	25
Agricultural research	15	+15*	30	+2	32
Medical research	6	+14*	20	+1	21
Nuclear research	3	+1	4	–	4
Industrial research	7	+15*	22	+3	25
Environmental research	2	+13*	15	+1	16
Total	70		158		
Increase		88		40	
Grand total			158	40	198

* Indicates bodies with rapid increase.
Source: Forje (1989), p. 52.

the outcome of the project is likely to suffer accordingly. If one considers the limited degree of indigenous technological capabilities as the 'trait' in question, it becomes possible to view the behaviour described in this section in terms of an excessive reliance on 'trait-taking' in public sector projects. This propensity may have permitted investment and output to increase more rapidly than if 'trait-making' had featured more prominently, but the result in the longer run was certainly to perpetuate the initially limited supply of domestic technological capabilities.

What seems just as clear is that the proliferation of science and technology institutions in Sub-Saharan Africa during the 1970s and early 1980s (shown in Table 1.9) had little or no influence over this generally disappointing outcome. One might have expected these institutions to exert an influence on the technological behaviour of public enterprises, which, after all, also belong to the public sector and ought therefore to have been relatively accessible to any such influence. Yet, for a variety of different reasons far too numerous to review here, the impact of science and technology institutions on indigenous technological capabilities (and for that matter technological behaviour in general) has been remarkably slight.[45]

CONCLUSIONS

This chapter has emphasized on the one hand, the extent to which public investment in Africa has increasingly come to depend on external sources of finance, and on the other the tendency for this investment to take systematic (and from most economic points of view undesirable) technological forms. We have not yet, however, considered any of the causal connections between these phenomena. Nor, relatedly, have we discussed particular cases that challenge the stylized account of technological behaviour that this chapter has presented. It is to these various issues that most of the following chapters are devoted.

Notes

1. Steel and Evans (1984), p. 60.
2. Hyden (1983), p. 96.
3. Clark (1978), p. 178.
4. Ibid.
5. Grayson (1973), p. 481.
6. Ibid., p. 486.
7. Mytelka (1984), p. 158.
8. Ibid.
9. Gulhati and Datta (1983).
10. Gulhati and Datta (1983) provide evidence for 1977–9 that the share of public investment in total investment was 93 per cent. More generally, they find that 'At the end of the 1970s, there was scarcely any country in our sample in which the share of public investment fell short of 40% of total investment in the formal or monetized sector and in several, this share was much higher'. (p. 33).
11. The projects were Mwanza Textiles, Morogoro Industrial Complex, Morogoro Textiles and Mufindi Pulp and Paper. The first of these projects is discussed briefly in the following chapter and in James (1983); The Morogoro shoe factory is mentioned in Lall (1987) and James (1983); The Pulp and Paper project is considered in quite some detail by Skarstein and Wangwe (1986).
12. World Bank loans to DFCs in the Third World began after 1950. By 1979, the Bank was supporting over 100 DFCs with total commitments amounting to more than US$6 billion (most of which was lent to the manufacturing sector). Unlike its direct investments (such as those mentioned in the previous note), the Bank does not appraise the individual projects financed through DFCs. Instead, only the appraisals conducted by the DFCs themselves are appraised. For further discussion of the Bank's relationships with DFCs see World Bank (1982).
13. Babai (1988).
14. Ibid., p. 260.
15. Ibid.

16. Lall (1992), p. 109.
17. Steel (1972). The DRC measures the domestic resource cost needed to save or earn one unit of value added (at border prices). A low value of the DRC indicates efficiency in the firm's activities; a high value just the opposite.
18. World Bank (1984), p. 67.
19. World Bank (1987), p. xi.
20. Bagachwa and Stewart (1992), p. 161.
21. World Bank (1984), p. 67.
22. For details see Millward (1988).
23. A useful diagrammatic explanation of this distinction is provided in Gillis et al. (1992).
24. Perkins (1980).
25. Bagachwa (1992).
26. Williams (1976) and Clark (1978).
27. Pack (1982).
28. Ibid., p. 6.
29. Khan and Thorbecke (1988). See also James and Khan (1993).
30. Stewart (1981), p. 80. A useful recent review of the growing literature on technological capabilities is provided in Lall (1992).
31. Akpakpan (1986).
32. Wangwe (1986), p. 15; emphasis added.
33. Mytelka (1992).
34. Grayson (1973); Adei (1990).
35. Juma (1991).
36. Barker et al. (1986).
37. Ohiorhenuan and Poloamina (1992).
38. Mytelka (1985), p. 87.
39. Barker et al. (1986), p. 121.
40. Adei (1990), p. 1053.
41. Ohiorhenuan and Poloamina (1992), p. 316. This study also finds that 'The industrialisation model adopted by publicly owned enterprises emphasises mass production of a standardised good and employment, rather than basic technology acquisition' (p. 304).
42. Akpakpan (1986).
43. Hirschman (1967).
44. Ibid., p. 131.
45. For reviews of the performance of science and technology institutions in Sub-Saharan Africa the reader is referred to Vitta (1990), Goka, Mihyo and Osunbor (1990), Unesco (1987), Forje (1989).

References

S. Adei (1990). 'Technological Capacity and Aborted Industrialization in Ghana: The Case of Bonsa Tyre Company', *World Development*, vol. 18(11).

J. Ahiakpor (1986). 'The Capital Intensity of Foreign, Private Local and State Owned Firms in a Less Developed Country', *Journal of Development Economics*, vol. 20.

E. Akpakpan (1986). 'Acquisition of Foreign Technology: A Case Study of Modern Brewing in Nigeria', *Development and Change*, vol. 17(4).

D. Babai (1988). 'The World Bank and the IMF: Rolling Back the State or Backing its Role', in R. Vernon (ed.), *The Promise of Privatization: A Challenge for US Policy*, Council on Foreign Relations.

M.S.D. Bagachwa (1992). 'Choice of Technology in Small and Large Firms: Grain Milling in Tanzania', *World Development*, vol. 20(1).

M.S.D. Bagachwa and F. Stewart (1992). 'Rural Industries and Rural Linkages in SubSaharan Africa: A Survey', in F. Stewart, S. Lall and S. Wangwe (eds), *Alternative Development Strategies in SubSaharan Africa*, Macmillan.

C.E. Barker, M.R. Bhagavan, P. v. Mitschke Collande and D.V. Wield (1986). *African Industrialisation*, Gower.

W.E. Clark (1978). *Socialist Development and Public Investment in Tanzania*, University of Toronto Press.

J. Forje (1989). *Science and Technology in Africa*, Longman.

M. Gillis, D. Perkins, M. Roemer and D. Snodgrass (eds) (1992). *Economics of Development*, 3rd edn, W.W. Norton.

A. Goka, P. Mihyo and O. Osunbor (1990). *Performance Review of Institutions for Technology Policy in Ghana, Nigeria and Tanzania*, IDRC, Manuscript Report 241e.

L. Grayson (1973). 'The Role of Suppliers' Credits in the Industrialization of Ghana', *Economic Development and Cultural Change*.

R. Gulhati and G. Datta (1983). 'Capital Accumulation in Eastern and Southern Africa: A Decade of Setbacks', World Bank Staff Working Paper, no. 562.

A. Hirschman (1967). *Development Projects Observed*, The Brookings Foundation.

P. Hopcraft and J. Oguttu (1982). 'Parastatal Development Agencies and their Relationship with the Private Sector', Institute of Development Studies, University of Nairobi, Occasional Paper no. 39.

G. Hyden (1983). *No Shortcuts to Progress*, Heinemann.

J. James (1983). 'Bureaucratic, Engineering and Economic Men: Decision-Making for Technology in Tanzania's State-Owned Enterprises', World Employment Programme Research, Working Paper no. 125, ILO, Geneva.

J. James and H. Khan (1993). 'The Employment Effects of an Income Redistribution in Developing Countries', *World Development*, vol. 21(5).

C. Juma (1991). 'Investment Strategy and Technology Policy: Lessons from Power-Alcohol Development', in P. Coughlin and G. Ikiara (eds), *Kenya's Industrialization Dilemma*, Heinemann.

H. Khan and E. Thorbecke (1988). *Macroeconomic Effects and Diffusion of Alternative Technologies within a Social Accounting Matrix Framework*, Gower.

T. Killick (1978). *Development Economics in Action*, Heinemann.

S. Lall (1987). 'Long Term Perspectives on Sub-Saharan Africa', Background Paper, Industry Division, World Bank.

S. Lall (1992). 'Technological Capabilities and Industrialisation', *World Development*, vol. 20(2).

R. Millward (1988). 'Measured Sources of Inefficiency in the Performance of Private and Public Enterprises in LDCs', in P. Cook and C. Kirkpatrick (eds), *Privatisation in Less Developed Countries*, Harvester Wheatsheaf.

R. Mukandala (1988). *The Political Economy of Parastatal Enterprise in Tanzania and Botswana*, PhD thesis, University of California, Berkeley.

L. Mytelka (1984). 'Foreign Business and Economic Development in the Ivory Coast', in I. Zartman (ed.), *The Political Economy of Ivory Coast*, Praeger.

L. Mytelka (1985). 'Stimulating Effective Technology Transfer: the Case of Textiles in Africa', in N. Rosenberg and C. Frischtak (eds), *International Technology Transfer: Concepts, Measures and Comparisons*, Praeger.

L. Mytelka (1992). 'Ivorian Industry at the Cross-Roads', in F. Stewart, S. Lall and S. Wangwe (eds), *Alternative Development Strategies in SubSaharan Africa*, Macmillan.

J. Ohiorhenuan and I. Poloamina (1992). 'Building Indigenous Capacity in African Industry', in F. Stewart, S. Lall and S. Wangwe (eds), *Alternative Development Strategies in SubSaharan Africa*, Macmillan.

H. Pack (1982). 'The Aggregate Implications of Factor Substitution in Industrial Processes', *Journal of Development Economics*, vol. 11(1).

F. Perkins (1980). *Technological Choice, Industrialisation and Development: The Case of Tanzania*, PhD thesis, University of Sussex.

F. Perkins (1983). 'Technology Choice, Industrialization and Development: The Case of Tanzania', *Journal of Development Studies*, vol. 19.

R. Skarstein and S. Wangwe (1986). *Industrial Development in Tanzania: Some Critical Issues*, Tanzania Publishing House.

W. Steel (1972). 'Import Substitution and Excess Capacity in Ghana', *Oxford Economic Papers*, vol. 24(1).

W. Steel and J. Evans (1984). 'Industrialization in Sub-Saharan Africa: Strategies and Performance', World Bank Technical Paper no. 25.

F. Stewart (1981). 'International Technology Transfer: Issues and Policy Options', in P. Streeten and R. Jolly (eds), *Recent Issues in World Development*, Pergamon.

D. Swanson and T. Wolde-Semait (1989). 'Africa's Public Enterprise Sector and Evidence of Reforms', World Bank Technical Paper, no. 95.

Unesco (1987). *Comparative Study on the National Science and Technology Policy-Making Bodies in the Countries of Eastern and Southern Africa*, Paris.

P. Vitta (1990). 'Technology Policy in Sub-Saharan Africa: Why the Dream Remains Unfulfilled', *World Development*, vol. 18(11).

S. Wangwe (1983). 'Industrialization in Tanzania', mimeo.

S. Wangwe (1986). 'Technology Imports, Technological Learning and Self-Reliance in Tanzania', World Employment Programme Research, Working Paper no. 165, ILO, Geneva.

S. Wangwe (1992). 'Building Indigenous Technological Capacity: A Study of Selected Industries in Tanzania', in F. Stewart, S. Lall and S. Wangwe (eds), *Alternative Development Strategies in SubSaharan Africa*, Macmillan.

D. Williams (1976). *Choice of Technology and National Planning: The Case of Tanzania*, PhD thesis, School of Business Administration, Harvard University.

World Bank (1982). *Technology and Science in World Bank Operations*, Washington, DC.

World Bank (1984). *Zambia: Industrial Policy and Performance*, Industry Dept, Industrial Strategy and Policy Division.

World Bank (1987). *Tanzania: An Agenda for Industrial Recovery*, Africa Region, Industrial Development and Finance Division.

2 The Bureaucratic Choice of Techniques: Textiles in Tanzania

This chapter is important for the light it throws on technological behaviour in the public sector of one African country which cannot adequately be explained by conventional arguments. In modified form, moreover, the theoretical framework of this chapter also finds expression later in the book.

THE PUZZLE: PRONOUNCED TECHNOLOGICAL VARIATIONS IN THE PUBLIC SECTOR

That the Tanzanian state should have more or less simultaneously established two large-scale integrated textile plants in the late 1960s is not particularly surprising. After all, following the Arusha Declaration of 1967, the leaders of that country were anxious to become self-reliant in key industries within a relatively short period of time. What was remarkable, though, was the extent of the divergence in factor intensities between the two plants, a divergence that far exceeded the then existing range of technologies in the public sector. One plant, Mwanza Textiles, employed a range of what were the most modern and automated technologies imported from Western Europe, while the other, the Friendship Mill, operated with Chinese equipment, some of which was designed in the 1930s.[1]

Table 2.1 shows that the latter was significantly more labour-intensive than the former; the table also indicates that the Chinese factory was more profitable than the plant which relied on relatively automated techniques of production. More generally, in the textile industry as a whole, a fairly close (inverse) correspondence between the degree of labour intensity and the cost of production was reported by Williams in his detailed 1976 study.[2] (At shadow prices, as one would expect, this correspondence became even more pronounced, increasing the relative profitability of the labour-intensive techniques.) Some years later, in the early 1980s, a comparison of the same technologies on the basis of a different set of productivity and cost measures led Frances Perkins to very similar conclusions.[3]

Most recently, a very detailed study of the textile sector by de Valk shows that Friendship has performed better overall than any other state-owned

Table 2.1 Friendship and Mwanza textile mills in Tanzania

Capital cost up to 1969 (million shillings)	61.5	106.5
Production of woven fabrics in 1975 (million linear metres)	24.0	22.5
Number of employees in 1975	5057	2486
Profit in 1975 (million shillings)	2.8	2.3
Cost of carded cotton in 1973 (shillings per tonne)	2512	2910
Labour hours per tonne of carded cotton in 1973	247	98

Source: Coulson (1982), p. 282.

textile plant in Tanzania;[4] in fact it performed better than most privately owned enterprises as well. One cannot of course attribute this outcome entirely to technological factors, since many other influences also affect firms' performance. We do know however, from the previous chapter, that reductions in the degree of capital and import intensity tend significantly to increase the efficiency of firms (as measured by domestic resource cost) in the industrial sector in Tanzania. It is plausible to argue therefore that the low domestic resource cost calculations for Friendship (which are less than one over a wide range of the firm's output) are, at least to some extent, attributable to the technology that was chosen. Mwanza, on the other hand, performs poorly on all indicators of performance, at least partly, one strongly suspects, on technological grounds.

The intriguing question to which all of this evidence gives rise is, How is it possible for the state simultaneously in the same sector to maintain such seemingly divergent technological preferences? The answer to this question according to Williams is certainly not to be found within the received choice of technique literature.

The Limits of Conventional Explanations

Defined in terms of embodied characteristics, product differences have often been cited as a major explanation of observed differences in factor intensities among alternative production processes. Yet this factor did not appear to be decisive in accounting for the marked technological variations that Williams and Perkins both identified. On the contrary, as Williams himself put it in 1975:

Within the range of technology currently in operation in Tanzania, the research showed that, in the opinion of production managers, the product has very little influence on the type of equipment across a range of yarn construction. In effect, the mills could produce the same products at comparable quality. There were no grounds for assuming that any particular type of technology in the observed range was dictated by product characteristics.[5]

Another popular class of explanations in the choice of technology literature imputes various types of non-economic objectives to decision-makers in developing countries. Probably the best known of these is the 'engineering-man' hypothesis associated with Wells.[6] Stated briefly, his contention is that under non-competitive conditions, the preference of 'engineering man' for sophisticated automated technology overrides the concern of 'economic man' to minimize costs of production. The conditions in Tanzania's textile industry in the late 1960s were certainly imperfect enough to allow the engineers to hold sway but one would then expect a *uniform* bias in favour of modern, capital-intensive equipment. 'The engineering-man hypothesis would not predict that a single investor . . . would set up both capital-intensive and labour-intensive plants at the same time.'[7]

It bears emphasizing that exactly the same problem would beset any of the numerous other technology-related objective functions that have been advanced in the choice of technology literature. Hawrylyshyn[8] and Winston[9], for example, both suggest that for decision-makers in developing countries, the bias in favour of technological modernity frequently derives from an equation of this type of technology with a deeply felt sense of 'national pride'. In particular,

> If a manager follows economists' admonitions and invests in a new plant with simple, labour-intensive, uncomplicated techniques of production, he accepts his people's ethnic inferiority, casting it in steel and concrete. . . . To reject appropriate technology, to choose instead the sophisticated 'high technology' . . . is to assert the power, equality and technical competence of one's people.[10]

However, because one would expect its alleged bias to apply (more or less) *consistently* to decisions made by a particular agency of the state at a given point of time, this sociologically oriented explanation takes us no further than does engineering man in accounting for the marked technological differences described above. Following Williams, it is more promising to investigate instead the nature of *non-technology* related objective functions in the public sector.

THE BUREAUCRATIC CHOICE OF TECHNIQUES

The Primacy of the Budget in National Planning

What Williams refers to as the 'bureaucratic choice of techniques' is rooted to a considerable degree in the alleged failure of the planning system in Tanzania. In particular, far from becoming a system that has effectively translated the major goals of the leadership into concrete achievements (as was originally intended), planning in Tanzania has, in his view, effectively been reduced to what is entailed in the budgeting process. The Development Budget, that is to say,

> became, and remains, the focus of planning in Tanzania. In theory this process is the last step in planning and simply apportions the available resources to those ends which planning has determined. But, since the flow of information to planners is limited, and hence the knowledge of the planners is limited; and since numerous decisions which predetermine allocation are made outside the planning system, budgeting is the only form of planning that remains. In the face of reality the organizational procedures of the planners have moved steadily towards identity with allocation. The planning horizon is a year and plans are written round the budget. Influence on plans made by other institutions is exercised only through the budget.[11]

This disappointing, if not unusual, outcome seemed to reflect a wide variety of factors, not the least important of which (especially in the early years following the Arusha Declaration) was a marked shortage of the required planning skills. Above all, though, this outcome seemed to reflect bureaucratic politics. For in Tanzania, as elsewhere, planning is not simply an organizational or technical matter. It entails also a range of political factors, which arise in large part because of conflicts between the various agencies of the state. The outcomes of these conflicts, in turn, help to determine the extent to which planning goals are translated into real achievements. Much depends, as the Tanzanian example illustrates, on the influence that the Ministry of Planning is able to wield over other state institutions, for in that country, such efforts as were made by the planners seemed to bring them into continual conflict with the Treasury.[12] And in the political struggle that ensued, it was the power derived from being in control over financial resources – especially foreign sources of funds – that ultimately enabled the Treasury to prevail (indeed, in 1975 important sections of the planning function were themselves incorporated into divisions of the Treasury).[13]

Conflicts between various ministries also intruded on the planning process. For what was essentially at issue in these conflicts was the priority of the objectives that were being separately pursued by the ministries. And if they were to perform the role of adjudicators over these competing claims, the planners clearly needed the requisite political authority for the task.

> But the ministers often had access to the President on the basis of personal relations. Against this the planners had little strength, and what they had rested in the allocative powers (borrowed from Treasury) on the Development Budget.[14]

On account of some of these same failings, Tanzania is not alone among African countries in exhibiting only a very tenuous relationship between the planning and budgeting functions. Indeed, a 1989 report for the World Bank on the management of public expenditure in developing countries found it 'discouragingly true' that 'after great effort has been devoted to formulating an investment plan, the budget, even for the first year following the plan's adoption may bear little or no relation to it'.[15] Nowhere, it seems, has the problem been more acute than in Sierra Leone, as the following description of the relationship between planning and budgeting in that country all too clearly indicates:

> there is no evidence that the overall goals of the plan, sectoral objectives or inter-industry relations are taken into account in preparing the yearly development expenditure budget and in selecting the projects for the sectoral investment programs. Sometimes, a project is included for political reasons in disregard of considerations such as total costs, availability of funds and long run development objectives. In other cases, the availability of funds becomes the determining factor, regardless of the project's repercussions on other sectors, on the balance of payments and on the Government Budget.[16]

A very similar story can be told for Kenya, in spite of the existence in that country of a system of forward budgeting that is designed specifically to forge a link between the development plan and the annual budget.[17] The Kenyan case is described in more detail in the following chapter.

Budgeting and the Choice of Techniques

How, then, did the alleged failure of the planning system in Tanzania impinge on the technological behaviour of the state-owned enterprises in the industrial sector? In Williams's account, the answer to this question proceeds largely along the following lines.

Devoid as they were of any detailed guidance from the planners, 'The institutions in the parastatal environment were motivated by their own perceptions of the indicators of success'.[18] However, there was nothing in this environment to suggest to the parastatals that efficiency warranted anything more than passing concern. For, on the one hand, 'the future of both firms and their managers seemed assured no matter what kind of commercial performance they turned in',[19] partly it seems, because prices were invariably adjusted to cover costs of production. On the other hand,

> Lack of concern for cost and efficiency was exacerbated by the rapid turnaround of personnel in the bureaucracy. Even if low profitability could give rise to criticism, the manager who implemented a project would probably be far away before the true nature of an investment emerged from two or three years of 'excusable' start-up difficulties.[20]

And even then low profitability would not necessarily have focused attention on the major *technological* aspects of the project, unless clear signals in the form of major mechanical breakdowns and loss of production emerged.

Freed thus from any constraints on costs imposed by either the protected macro-economic environment or the national planning system, parastatal managers turned their attention instead to other objectives. Chief among these according to Williams was to implement projects as rapidly as possible. Exactly why this objective should have been deemed so important by managers is not made clear, though it seems to have been related to the 'implicit strategy' of the top leadership that had *speed* as its cornerstone. 'Speed was regarded as essential. Time was seen as being against a poor nation combating rising expectations and seeking to avoid the growth and entrenchment of a dominant domestic capitalist system. Visible achievements related to the goals were highly important. Self-reliance meant Tanzanian institutions starting more Tanzanian industries. . . .

Thus, the society's goal structure and implicit strategy . . . appeared to require *rapid implementation of projects.*'[21]

But the parastatal manager was constrained in this goal primarily by financial needs, that is, in identifying projects and getting them through the budgeting process. And this requirement, in turn, led to rather predictable behaviour on the part of managers. It led, above all, to a reinforcement of the tendency towards selecting *highly packaged projects*; a tendency that was already imposed on parastatals by skill constraints and by lack of information about technology and other markets.[22] For highly packaged projects usually also displayed distinct advantages not only in budgetary terms but also in terms of the ease and rapidity with which they could be implemented.[23] Indeed, 'a project package which included financing and which was "ready to go" would often be accepted with little question'.[24] On the other hand, 'The most difficult projects for which to gain acceptance were those which were to be locally funded, were "experimental" in that they proposed, for example, the use of unfamiliar technology.'[25] (Projects were also more likely to be approved if it could be demonstrated that binding international commitments had already been established on their behalf; refusal by the planners might then have led to embarrassment of the government.)[26]

Under these circumstances, concludes Williams, the 'choice of technique' is merely the by-product of a bureaucratic process, the 'fall-out' from the particular package of external finance and other project inputs that happen to be chosen. What matters in this process is not the value of one project in relation to another but rather the demands made by each of them on scarce equity finance. (In fact, any concern there may be with project viability in the process, so the argument goes, is usually confined to a single feasibility study and a very cursory evaluation thereof by the planning authorities.) In their turn,

> The parastatals responded to the bureaucratic realities by presenting projects in a manner best calculated to ensure acceptance. *Thus, a low-cost labour-intensive textile mill, completely financed outside the budget would be preferred to any mill for which more financing from the budget [was] required. But if circumstances were such that the more acceptable package available consisted of a high-cost capital intensive mill, then that would be chosen.*[27]

It is therefore to *changes* in the form taken by what were regarded as acceptable *projects* that one needs to look in explaining technological variations in the textile sector (and for that matter in other sectors of industry as well).[28] Consider first the relatively labour-intensive Friendship mill, mentioned earlier in connection with Table 2.1. Conceived originally as the result of high-level political contacts, this project was entirely financed (with respect to its investment costs) by a long-term interest-free loan to the Tanzanian government from the People's Republic of China. By virtue of being both externally financed and politically sanctioned, the project received the automatic approval that the bureaucratic choice of techniques would lead one to expect. And, as that model also predicts, the approval process did not entail any technological concerns on the part of the Tanzanian bureaucracy. The technology that was 'chosen' for the Friendship plant seemed to reflect, instead, the familiarity of the Chinese with older and relatively labour-intensive technological vintages. In fact, according to one source cited by Williams, the particular vintage used at Friendship was the most labour-intensive of the textile technologies then being produced in China.[29]

One year later, however, in the form of Mwanza Textiles, a very different (albeit still attractive) package of project characteristics presented itself. It was different, in the first place, because the external finance took the form of a supplier's credit rather than foreign aid. It was also different from the Friendship case in that the technological 'fall-out' from the chosen source of finance and related project inputs led to Western Europe (especially France) rather than to China, and partly in consequence to 'the latest and most automated equipment'[30] rather than to technology that at the time was much more dated. The inference here is not that technologies imported from the developed countries are *inevitably* more capital-intensive than those emanating from the developing countries themselves. The weaker claim, as argued in Chapters 3 and 5 below, is that there are general historical grounds for expecting some such association to exist. By failing to address himself to these historical factors, we contend that Williams has unduly confined the explanatory range of his bureaucratic choice of techniques.

What Williams does recognize, however, is that other seemingly paradoxical forms of technological behaviour by the Tanzanian state also seem amenable to the supply-oriented explanation that has just been applied to the case of textiles. In certain other sectors, that is to say, there are on the one hand, an equally wide range of factor intensities and on the other, just as little evidence to suggest that these variations actually reflect the technological preferences of the state.[31] Instead, as in textiles, the state's preferences tend to be defined over project *packages* and it is to the determinants of these that one needs to look for an explanation (along the lines that are suggested in the following three chapters). Far, therefore, from refuting the bureaucratic choice of techniques (as Kaplinsky has recently suggested[32]), the simultaneous use in the public sector of appropriate and inappropriate brickmaking technologies may simply reflect the fact that one project package was financed by an international appropriate technology institution while the other was not.[33]

CONCLUSIONS

This chapter has examined the notion of a bureaucratic choice of techniques as it was originally formulated and it has shown how this idea can throw light on technological behaviour in the public sector of one country, Tanzania, that seems to defy conventional explanations. Moreover, by drawing attention to the attractions of highly packaged projects, this chapter helps to explain the predominance of these types of projects in the public sector of Tanzania and other African countries (noted in Chapter 1). At the same time, however, we suggested that as originally formulated, the bureaucratic choice of techniques is lacking in several important respects. The following chapters, accordingly, seek to redress these deficiencies and thereby to increase our understanding of the technological behaviour of state institutions in Africa.

Notes

1. Williams (1976).
2. Ibid.
3. Perkins (1983).
4. de Valk (1992).
5. Williams (1975), p. 3.
6. Wells (1975).
7. Williams (1975), p. 7.
8. Hawrylyshyn (1978).
9. Winston (1979).
10. Ibid., p. 840.
11. Williams (1976), p. 158.
12. Ibid.
13. Ibid.
14. Ibid., p. 162.
15. Lacey (1989).
16. Ibid., p. 17.
17. Ibid., p. 17.
18. Williams (1976), pp. 162–3.

19. Ibid., p. 172.
20. Ibid. Accordingly, parastatals modified their decision-making processes in an attempt to reduce the risks of mechanical breakdowns. But in itself this did nothing to increase the weight that was accorded to questions of efficiency and factor prices.
21. Ibid., pp. 153, 154; emphasis added.
22. The parastatals, that is to say, used highly packaged projects to overcome their own managerial constraints. For the same reason, direct foreign investment is often attractive to those developing countries that are least well endowed with basic technological capabilities.
23. From this point of view, suppliers' credits and turnkey projects are especially appealing.
24. Williams (1976), p. 163.
25. Ibid.
26. Ibid.
27. Ibid., p. 165 (emphasis added).
28. Following Williams (1976) one should note however that during the 1970s, some efforts were made by the parastatals to loosen the influence of external finance on projects by appointing independent consultants. Evidently, therefore, some learning from the disappointing experiences of the 1960s had taken place. It is not clear though how much influence the new procedures actually had.
29. Ibid.
30. Perkins (1980), p. 272. In the Mwanza case, the high capital costs may also have been partly due to over-invoicing of the equipment. See Williams (1976).
31. Some of these cases are discussed in Chapter 3.
32. Kaplinsky (1990).
33. Not surprisingly, the appropriate technology institution made strenuous efforts to identify and develop a technology that was suited to small-scale, labour-intensive production. See Kaplinsky (1990). In the other project however no such efforts were apparent.

References

A. Coulson (1982). *Tanzania: A Political Economy*, Oxford University Press.

O. Hawrylyshyn (1978). 'Capital-Intensity Biases in Developing Country Technology Choice', *Journal of Development Economics*, vol. 5.

R. Kaplinsky (1990). *The Economies of Small*, IT Publications.

R. Lacey (1989). 'The Management of Public Expenditures: An Evolving Bank Approach', Policy, Planning and Research, WPS 46, World Bank.

F. Perkins (1980). *Technological Choice, Industrialisation and Development: the Case of Tanzania*, DPhil. thesis, University of Sussex.

F. Perkins (1983). 'Technology Choice, Industrialisation and Development Experiences in Tanzania', *Journal of Development Studies*, vol. 19.

P. de Valk (1992). *A General Framework for Evaluating the Performance of*

Textile Enterprises in LDC's, with an Application to Tanzania under Structural Adjustment, PhD thesis, Free University, Amsterdam.

L. Wells (1975). 'Economic Man and Engineering Man' in C.P. Timmer et al., *The Choice of Technology in Developing Countries*, Center for International Affairs, Harvard University.

D. Williams (1975). 'National Planning and the Choice of Technology: the Case of Textiles in Tanzania', Economic Research Bureau Paper no. 75.12, University of Dar-es-Salaam.

D. Williams (1976). *Choice of Technology and National Planning: the Case of Tanzania*, PhD thesis, Graduate School of Business Administration, Harvard University.

G. Winston (1979). 'The Appeal of Inappropriate Technologies: Self-Inflicted Wages, Ethnic Pride and Corruption', *World Development*, vol. 7(8/9).

3 The Irrelevance of Ideology in the Public Sector: A Comparison of Kenya and Tanzania

Chapter 1 drew attention to the similarity in the technological behaviour of public enterprises in the industrial sector of various African countries. This observation is striking because it does not conform to what one might have expected on the basis of the highly disparate development strategies that were adopted in countries such as Kenya and Tanzania. One might have expected instead that the capitalistic orientation of the former and the (African) socialism espoused in the latter would have produced a correspondingly diverse pattern of technological behaviour on the part of the public enterprises concerned. For, to the extent that these enterprises can be considered as being essentially an extension, or 'creature' of the state, their technological behaviour ought presumably to reflect the divergent goals of the countries they represent.

The purpose of this chapter, accordingly, is to explain these somewhat paradoxical findings using data obtained mostly from a detailed study of how technology decisions were actually made in a sample of publicly owned firms drawn from the manufacturing sectors of Kenya and Tanzania. It is contended that in general these decisions seem to be dominated by a single objective which the parastatal holding companies are able to pursue almost entirely unimpeded by countervailing pressures from either the macro-environment or from other agencies of government and that this objective predisposes managers to search for investment projects in which, through various mechanisms and with varying degrees of directness, developed country patterns of scale, factor intensity and product characteristics are invariably closely reproduced.

The first part of the chapter describes and compares the data that bear on the technological characteristics of public enterprise in the two countries. In the following section a framework is proposed to explain these observations and thereafter, in the third section, the framework is applied to the chosen sets of case-studies. The concluding fourth section considers the main policy implications of the findings.

THE TECHNOLOGICAL CHARACTERISTICS OF PUBLIC ENTERPRISES IN MANUFACTURING IN KENYA AND TANZANIA: AN OVERVIEW OF A COMMON PROBLEM

Reflecting perhaps their differential prominence in the manufacturing sectors of the two economies, parastatals in Tanzania have been subject to a far greater degree of analytical scrutiny than in Kenya. With respect to the technological aspects of the behaviour of these enterprises in particular, the two major studies of the Tanzanian experience – by Clark[1] and Perkins[2] – are substantially more extensive in coverage than the corresponding evidence for Kenya. As a result, the assertions based on these studies, shown juxtaposed in Table 3.1, differ in the degree to which they can be said to characterize accurately the manufacturing sector as a whole. But such comparative data as do exist strongly suggest the interesting paradox that is common to the experience of both countries, namely, that the essentially *unplanned* character of the technological behaviour of public enterprises coincides with, and is indeed largely produced by, the *systematic* tendency for large-scale and (often inefficient) capital-intensive technologies to be associated with each other and with excessively sophisticated products. It is this seemingly paradoxical combination of features that appears to underlie the striking *similarity* of what has occurred in Kenya and Tanzania and to an understanding of which this chapter is designed to contribute, beginning, in the following section, with the elaboration of an analytical framework.

THE ANALYTICAL FRAMEWORK

According to Gillis, the traditional view in economics is that managers of public enterprises are 'cosmic maximisers'. That is, they are 'motivated solely by a desire to maximise a clearly defined measure of social welfare, as defined by the parent government, under conditions of perfect information on all shadow prices, externalities and risk.'[3] It is apparent even from the brief description of the common problem presented in the previous section, however, that this idealized model (which assumes not only that government is able to provide the firm with an unambiguous set of goals but also that managers actually seek to maximize according to them) will not do. For in both Kenya and Tanzania what has to be explained is precisely why the (very similar) technological behaviour of public enterprises diverges so sharply from what appear to be many of the most important objectives of the respective governments. What is required instead, therefore, is an alternative theoretical framework, which, on the one hand, posits an alternative behavioural assumption at the level of the enterprise and, on the other, contains an alternative and more realistic model of the relationship between the firm and the agencies of government. It needs to be stressed that both of these modifications are necessary for an alternative framework, for even if public enterprises do seek to pursue goals different from those that have been articulated by government and which lead to the type of technological outcomes observed in Kenya and Tanzania, one is bound nevertheless to explain the relationship between the firms and the agencies of government that *enables* this to occur.

With respect to the behavioural aspect of an alternative framework, our point of departure is the view that conditions in both Kenya and Tanzania in the 1970s were basically such as to produce what was described in the previous chapter as the 'bureaucratic choice of techniques'. As it was originally formulated by David Williams, that approach emphasizes the role of large-scale, turnkey projects in meeting the requirements of rapidly raising finance for projects and for ensuring their rapid delivery and construction. Such projects, moreover, also have the advantage of providing economies of scale in terms of scarce managerial resources at headquarters.

But while the bureaucratic process has, therefore, a systematic bias associated with *large-scale* production (that is also highly packaged), it does not, according to Williams, also have a systematic factor-intensity bias. Rather, the factor-intensity of the technology that results from the bureaucratic-man process is thought to be 'random'.

Stated in this way it is clear that the hypothesis is capable of accounting for some, but by no means all, the observed characteristics of the choices made by parastatals in Kenya and Tanzania. In particular, it fails to explain not only the marked tendency to capital intensity that was noted above in relation to both countries, but also the sophisticated nature of the products that invariably appear to accompany this tendency. It is in our view possible, however, to modify the implications of the basic model in a direction that is more consistent with these empirical observations by incorporating into it a more realistic view of the finance bias that is imparted to the technological features of the selected projects. And central to this modification are the systematic historical links between the various dimensions of technology that Frances Stewart has described in relation to the experience of the developed countries.[4]

The first point that needs to be recognized is that most of the offers of foreign finance to Kenyan and Tanzanian parastatals seeking this means of expanding investment emanate from the developed market economies. Thus, despite its socialist orientation, even in Tanzania in the mid-1970s,

Table 3.1 A comparative summary of the technological characteristics of public enterprises in manufacturing

	Tanzania	*Kenya*
Degree of conformity to national development goals	'The failure of the post-Arusha companies to distinguish themselves significantly and favourably from the pre-Arusha companies points to one of the most important conclusions which can be made about the performance of the manufacturing parastatals, indeed about all parastatals. *There has been as yet no developmental innovation on the part of parastatals to make themselves more consistent with the Tanzanian ideology*. They have tended to want the most modern plants. Turn-key projects . . . have not been uncommon.' (Clark, 1978, p. 140)	'This study suggests that the parastatal investment companies have done little or nothing to push the manufacturing sector in the directions specified by the policy statements of government. Rather, they appear to have *aided and abetted some of the least appropriate features of the Kenyan industrial structure*.' (Hopcraft and Oguttu, 1982, p. 193)
	'In general the observed pattern of technological choice . . . only reflects to a limited degree Tanzania's articulated development goals.' (Perkins, 1980, p. 408)	'While too large a proportion of the development budget has been diverted to investments, the balance of that budget has not been effectively utilized either. The pattern of development outlined in Development Plans has often been ignored in practice in favour of *unplanned* projects.' (Ndegwa, 1982, p. 14)
Relation to the planning system	'The present system of investment decision-making in the parastatal sector is essentially one of *unplanned socialism*. . . . The results bear only an incidental relation to any formal development plan or to the proclaimed goals of industrial development.' (IBRD, 1977a, p. 125)	'Parastatals have played a large and increasingly important role in Kenya's life in recent years. Growth of this sector has been, in many respects, *unplanned and uncoordinated*.' (IBRD, 1983, p. 105)

Technological characteristics of public enterprises	'Despite the rhetoric, Tanzania's industrialization programme has, in general, promoted the establishment of enterprises using large-scale capital-intensive, often technically, and almost invariably economically inefficient techniques.' (Perkins, 1983, p. 231)	'The parastatal client firms tend to be large, capital intensive, import intensive and almost exclusively oriented toward a protected, over priced local market.' (Hopcraft and Oguttu, 1982, p. 193)
	'almost one-third of parastatals have chosen to manufacture products . . . for which a simpler, less technologically specified substitute exists.' (Perkins, 1983, p. 226)	'Despite the emphasis in policy statements on assistance to small, African owned and export-oriented firms, firms with government participation tend to be large (half of all Kenya's industrial firms employing more than 200 people have government financial participation).' (IBRD, 1983, p. 85)
	'There is a strong tendency for parastatals to adopt the most advanced techniques, and build the most modern buildings. In manufacturing this tendency is exhibited as an inclination towards large-scale, capital-intensive, "transformation" projects . . . There has been very little change in the orientation of the parastatal sector since the Arusha Declaration . . . There has been no significant development of small-scale rural industries.' (Clark, 1978, p. 125)	'Examples of unsound and poorly controlled investments can readily be found in such areas of activity as fertilizer, sugar, textiles and power alcohol. . . . [which have] resulted in uncontrolled cost escalations, inefficient technologies and unprofitable enterprises.' (Ndegwa, 1982, p. 42)
Relation to foreign finance	'Public investment today remains about as *heavily dependent* on foreign funds as it did ten years ago. About one-half of public investment has been foreign financed.' (Clark, 1978, p. 173)	'A major characteristic of most of the parastatals in Kenya is that they *depend heavily* on foreign financing and management advice.' (Oyugi, 1982, p. 64)

'Almost 35 per cent of Tanzania's assistance is in the form of bilateral aid from Western capitalist countries, while another third comes from the World Bank.'[5] (And in Kenya, not surprisingly, aid from these same sources comprises a much higher, almost exhaustive, percentage of the total.[6]) Secondly, the 'fall-out' (to use Williams's phrase) from this type of finance normally produces a combination of technological characteristics that is not at all *random* but has rather evolved *systematically* out of the particular socio-economic circumstances prevailing in the developed countries with which the finance is associated.

Stewart was the first to elaborate the view of a particular technology as being a 'vector' of characteristics (comprising, among other things, the nature of the product and the scale of production) and that changes in these characteristics in the developed countries (where almost all technical change occurs) are mutually interrelated. On the one hand, 'As incomes rise, wages rise, making existing technology uneconomic compared with labour-saving innovations. Rising incomes allow greater expenditure on investment per employee.'[7] On the other, 'Technical change in products in developed countries occurs in line with the rising incomes (and is partly responsible for those rising incomes), so that the balance of characteristics offered by new products corresponds to the changing demands of consumers as their incomes rise. The rising incomes have the effect of shifting demand towards different products with more sophisticated, labour-saving, higher-quality, etc. characteristics.'[8] Moreover, since an increase in the size of the market is another consequence of the rise in average incomes, there is also a tendency for the scale of output to increase with economic growth. Indeed, 'Historically, in the advanced countries the typical size of plant has grown faster than the market as a whole.'[9]

In so far, therefore, as the argument of the two previous paragraphs has validity, the finance-bias that defines the bureaucratic-man hypothesis has implications not only for the scale of output, but also for the factor-intensity of technology and the nature of the product. In this way, the behavioural implications of the theory become more consistent with the empirical observations described in Table 3.1. But even the reformulated version of the bureaucratic-man mechanism does not adequately describe the nature of the relationship between the parastatals and the agencies of government that *permits* the pursuit of a managerial goal, the technological implications of which, are, for the most part inimical to the proclaimed goals of development. It is to a general discussion of this issue that the remaining part of the conceptual analysis is accordingly devoted.

Apart from specific problems that vary from one country to another, there is a set of intrinsic difficulties in seeking to ensure that public enterprises are effectively used for implementing government policies. Much of this inherent difficulty can be explained in terms of agent-principal theory.

Raymond Vernon has expressed the basic problem in this branch of theory when he asks: 'How does the principal ensure that the agent acting for him responds to the same information and the same congeries of objectives as the principal would do if acting on his own behalf?'[10] While this problem pervades, by definition, all organizations in which agents act on behalf of principals, its severity depends largely on the degree to which the behaviour of the former can be made accountable to the latter. For without an adequate system of accountability, control of the agent by his principal becomes extremely difficult.

Howard Raiffa has correctly observed that the problems of accountability and control are likely to be much less acute in privately than in publicly owned firms, for in the former, 'Managers usually have a bottom-line figure that holds them accountable to some extent. In private enterprise, the profit motive is strong and serves as a sieve through which gross incompetents are weeded out and others rewarded.'[11] In public enterprises, however, as is well known, the objectives are usually far more diffuse, since these firms are formed to fulfil a variety of different functions.

In itself, however, the diversity of objectives of the public enterprise creates no particular problem for the accountability and control system. If, as is assumed in the 'cosmic maximization' model described earlier in this section, the numerous objectives of the enterprise are combined into an unambiguous objective function, managers could be judged, and held accountable, according to this function in the same way that the private enterprise manager is held accountable in terms of profits. In practice, however, an unambiguous objective function is rarely, if ever, presented to the managers of the public enterprise and the multiplicity of goals consequently creates severe problems of enterprise accountability and control.

The difficulties involved in aggregating multiple objectives into an operational composite index have been lucidly described by Raiffa in relation to the chief executive officer (the principal) of a public enterprise, whose problem it is to communicate the multiple conflicting objectives of the firm to his agents in such a way that they act as the principal would in the same situation. The dilemma facing the chief executive is this. On the one hand, if he is to be able to control his agents effectively he will need to formalize the trade-offs between the conflicting objectives of the firm, for only with such formalization can an unambiguous objective function be derived. On the other hand, because there is unlikely to be any kind of consensus among the board members about the various trade-offs, whatever formalization of these he enunciates will get him into 'political trouble' with at least some members of the board. 'Thus, the chief executive officer is in an uncomfortable squeeze: he is damned if he formalises his trade-offs and damned if he does not.'[12]

Raiffa's example can be readily extended to the case in which 'the government' is the principal and the public enterprise its agent, for the problem of distilling from the various agencies of government a comprehensive set of trade-offs among conflicting goals is likely to be no less serious than it is for the board or directors in the firm. Indeed, Vernon points to the 'disconcerting fact that, where conflicting and mutually inconsistent goals seem to exist, *politicians may find it undesirable – even dangerous – to try to clarify the ambiguity*.'[13]

The political difficulties of formalizing trade-offs (and consequently of controlling public enterprises effectively) may perhaps be especially severe in open, participatory democracies in which there is a diffuse set of opposing interest groups. Somewhat fewer difficulties may be predicted in less participatory regimes, where political power is more highly concentrated among a small group of decision-makers. But even in such cases, a quite considerable amount of disagreement about fundamental trade-offs among development goals can still exist.

What we have shown so far in this section is that there are some inherent (and seemingly often intractable) difficulties in making public enterprise conform to national goals. These difficulties often confer substantial autonomy on the enterprises. But there are also measures that may be *actively* pursued by managers to further their autonomy. The degree to which managers are successful in these endeavours is a further determinant of the likelihood that the outcomes of the behaviour of public enterprises will be at variance with what is intended by government. According to Aharoni, a variety of variables bear on the ability of the manager to increase his autonomy.[14] Among the most important of these are finance, the legal organization of the firm and the efficacy of the control functions exercised by government.

APPLICATION TO KENYA AND TANZANIA

It follows from the foregoing discussion that in order to apply the analytical framework to Kenya and Tanzania we shall need to concern ourselves with three major sets of issues in these countries, namely:

(a) the nature of the conditions that give rise to the operation of the bureaucratic-man mechanism,

(b) the technological 'fall-out' from this mechanism, i.e. the manner in which factor-intensity, scale of output and type of product are derived from the financial and other inputs embodied in the sample projects, and
(c) the relationships between the parastatals and relevant agencies of government that bear (theoretically and in practice) on this 'fall-out' process.

The Common Origins of the Bureaucratic Mechanism

It will be recalled that bureaucratic man has as his main objective the 'starting-up' of as many projects as possible. For in this focus is said to be 'the way to survive and prosper'. But quite what set of pressures operate to produce this sort of managerial motivation is not at all clear from Williams's statement of the hypothesis. In this part of the chapter it is argued tentatively that the source of these pressures in both Kenya and Tanzania can be traced to the extent of, and *rapidity* with which, output increases were sought (especially in key sectors) and the ease with which this objective of government could be assumed by managers of the parastatals. It will be argued, furthermore, that the goal of rapid (and sizeable) increments in output could be pursued by managers in both countries without any significant degree of pressure for cost-minimization arising from the macro-economic environment.

The juxtaposed entries in Table 3.2a, dealing with national goals, and goals for the textile and sugar industries, lend support to the view that a common desire for rapid growth in domestic output – of wage goods in particular – can readily be discerned from a study of official and other documents. (Emphases have been added in the tables that follow.) In both Kenya and Tanzania this desire would appear to have been fundamentally shaped by both political and economic factors – the former because of the significance that was apparently attributed to meeting the excess demand for many wage goods (see for example Barclay's view of this for the Kenyan sugar industry contained in Table 3.2a), and the latter because of the foreign exchange crisis that beset these countries during much of the decade. (In the Tanzanian case it should be noted that there was also a distinctively ideological basis for the quest for self-reliance in 'basic goods' that formed an essential part of the so-called basic industry strategy.[15])

The admittedly rather sketchy evidence presented in Table 3.2a suggests that these political and economic imperatives underlying the 'national' goal for industries such as textiles and sugar translated themselves, at the level of the parastatal holding companies and development finance

Table 3.2a The industrial objectives of 'government' in the 1970s and early 1980s

	Tanzania	*Kenya*
National goals	'The major objective . . . is to achieve a greater degree of economic self-sufficiency.' (Third Five Year Plan, p. 43) 'The present plan has placed a high priority on the satisfaction of domestic basic needs. Thus the production of food, textiles, shoes and construction materials will have particular emphasis.' (Third Five Year Plan, p. 43) 'The strategy puts a premium on the *earliest* feasible development of a basic industry.' (Roemer et al., 1976, p. 269) 'New projects will have to be consistent with the basic industrial strategy.' (Third Five Year Plan, p. 47)	'The overriding industrial strategy of the Government – that is to *accelerate* industrial growth.' (Second Five Year Plan, p. 326) 'The target [for 1974] is an increase in output of over 70 per cent compared with 1967.' (Second Five Year Plan, p. 12) 'The attainment of the income and employment targets in this Plan depends to a significant degree on the country obtaining physical resources from abroad to supplement its own effort. Without making use of these physical resources . . . the country cannot develop as *rapidly* as planned. The Government will therefore endeavour to obtain from abroad the resources necessary to achieve the national goals.' (Second Five Year Plan, pp. 329–30) 'The crucial point in the drive for *accelerated* industrial development is that highest priority be accorded to projects which make the greatest contribution to improving Kenya's balance of payments position.' (Annual Reports and Accounts of the Industrial Development Bank, 1974, p. 9)
The sugar industry	'The national goal is to achieve self-sufficiency in sugar production at *the earliest opportunity*, sustaining self-sufficiency at all times and resources and economics permitting produce surplus sugar for export.' (Sugar Development Corporation, Annual Report and Accounts, 1976–7, p. 33) '*Rapid* progress in this area is required to satisfy domestic demand.' (Third Five Year Plan, p. 46)	'Bearing in mind the sugar production deficit and the need for foreign exchange to purchase additional sugar for the country's consumption requirements, the Government requested funds . . . for a study to be made of the industry and the development of a programme for the country to become self-sufficient if economically possible *within a period of 10 years*.' (Government of Kenya, Kenya Sugar Industry, *Expansion Study*, vol. V, Nyanza Sugar Belt Improvement Project, Tate and Lyle, May 1976, p. 8)

		'Kenya's commitment to meeting the consumer demand for sugar by attaining self-sufficiency in its production. The latter theme has been consistently stated as central to agricultural and industrial policy during recent years. Satisfaction of consumer demand has remained a high political priority for the Government.' (Barclay, 1977, pp. 61–2)
The textile industry	'In order to satisfy projected local and export demand in 1985, it will be necessary to produce 220 million square metres of cloth. To achieve this *four mills must be established between 1977/78 and 1980/81.*' (Third Five Year Plan, p. 47)	'The Government attaches great importance to the *rapid* development of the textile industry.' (from the correspondence files of the Rift Valley Textile Project) 'Our keen determination to move with utmost *speed* [to complete the project] as desired by the Kenyan authorities.' (from the correspondence files of the Rift Valley Textile Project)
The parastatal holding companies, development banks, and new projects	'NDC's [the National Development Corporation's] interest is not to run companies for the purpose of profit maximisation, but rather to implement projects in certain sectors. In fact, NDC seems to be evaluated by the *number* of projects it implements . . . *The unofficial goal of NDC as expressed by* some of its employees, to push through as *many* projects as possible.' (Quoted in O'Brien, 1982, p. 22) 'The management [of NDC] is anxious to show success. Lever Brothers will come and erect a detergent factory in a year. . . . Despite the poor economics of such a project from a developmental point of view, NDC appears more successful than if it had spent the year trying to organize Ujamaa village work shops.' (Clark, 1978, p. 189)	'One of IDB's [the Industrial Development Bank's] important aims has been to attract investment funds from abroad.' (Annual Report and Accounts, 1977, Industrial Development Bank, p. 13) 'Since December 1974 total investments approved . . . have grown by an average of nearly 25% a year . . . The *total number of projects* investigated and considered for investment has increased even more *rapidly* because of IDB's search for new business.' (Annual Report and Accounts, 1977, Industrial Development Bank, p. 12)

Table 3.2b

Tanzania			Kenya		
NDC investments, 1965–82			*Development Finance Company of Kenya (DFCK) involvement in projects, 1965–83*		
Year	*Investment*	*New investment*	*Year*	*No. of projects with DFCK equity/loans*	*Matured projects*
1.1.1965	23.9	n.a			
1965	102.7	78.8	1965	10	–
1966	117.0	14.3	1966	14	–
1967	211.0	94.1	1967	21	–
1968	311.0	100.0	1968	21	–
1969	260.7	43.6	1969	29	–
1970	333.5	87.0	1970	31	–
1971	429.7	130.2	1971	38	–
1972	450.3	20.6	1972	41	–
1973	336.5	11.9	1973	46	–
1974	281.2	n.a	1974	49	2
1975	199.1	n.a	1975	53	–
1976	239.3	40.3	1976	58	3
1977	290.3	103.6	1977	64	2
1978	415.8	125.5	1978	83	2
1979	194.1	–	1979	100	1
1980	260.7	127.7	1980	119	1
1981	492.4	231.7	1981	141	3
1982	495.6	3.2	1982	145	2
			1983	156	3
			Total	1219	19

Note: 1980–2 figures exclude the Southern Paper Mills Company Ltd, Tanzania.
Sources: Mukandala (1988) p. 316; Ikiara (1991) p. 249.

companies (in which important decisions regarding manufacturing subsidiaries are mostly made), into a predisposition to initiate a large number of projects (as for example in the NDC in Tanzania and the DFCK in Kenya). The manner of this translation as well as its effects derives from the facts that maximizing current output is an objective that is easily measured (and in terms of which, therefore, kudos can be unambiguously earned) and that it is also a goal that can be rapidly achieved (as the quotation for Tanzania in Table 3.2a suggests) on the basis of foreign-financed projects – *the more so given that, as Table 3.3 clearly indicates, pressures from the macro-environment to minimize costs were largely absent from both countries.* This common absence of pressure on costs stemmed in large measure from government interventions that had the effect, at the level of the enterprise, of breaking the link between economic efficiency and profitability.

Of course, maximization of output could be achieved by the fuller use of existing projects and to some extent this method was used by parastatal managers (e.g. in the attempted reduction of excess capacity). But given the often severe foreign exchange constraints on the efficient operation of these enterprises and the difficulty of raising foreign capital for the purpose of achieving a fuller utilization of existing resources – foreign aid for example is usually not available for repairs and spare parts – output maximization tended to translate in practice into a concern with *new* projects.

In brief, then, what is being suggested here is that the parastatal holding companies became strongly inclined to search for foreign finance (as was in fact explicitly stated by the Kenyan Industrial Development Bank in one of the quotations in Table 3.2) and highly receptive to offers of such finance. And in practice this usually meant finance from the developed market economies for two main reasons. First, it was this type of finance that even in Tanzania tended to be most abundant, for, as Clark has correctly observed, 'The socialist overtones, far from repelling western donors, attracted them. . . . What it meant to most donors was that the elite in Tanzania, because of the leadership code, would not enrich themselves on foreign aid, and would in fact strive to enact programmes to better the mass of the population. The Nordic countries, Canada, and the World Bank all found this a striking contrast to the situation in most of the Third World.'[16] The second reason has to do with the relative *rapidity* (and certainty) of the output increases that are associated with Western finance. In particular, 'Western firms offer *quick*, reliable delivery of capital goods. To buy from countries like China *takes time*, and involves a degree of uncertainty.'[17] (Perhaps a third reason is that developed-country suppliers

Table 3.3 The macro-environment and pressures for cost-consciousness in the choice of technology

Tanzania	*Kenya*
'Both product and factor markets in Tanzania are nominally under the control of planners. In both areas several aspects of the competitive model have been replaced with devices which operate in quite different directions from market forces. The price-control system, for example, is in many cases based on a cost-plus formulation which both shelters the inefficient and gives only a weak incentive to become more efficient. . . . The incentive structure to which the parastatal manager responds offers little in the way of personal financial rewards and, in any case, may focus on surpluses which are more related to windfall gains from the pricing system than to productive efficiency.' (Williams, 1975, p. 7)	'Financial participation by Government is the best guarantee obtainable that a firm will make high and secure financial profits *regardless of its economic efficiency or of its international competitiveness*.' (Hopcraft, 1979, p. 18) 'Firms with Government participation use it to make Government behave in the firm's interest. The complete range of special concessions, licenses, guarantees, and the whole edifice of protective measures, are far more accessible to the firm that has Government financial participation than to one that does not.' (IBRD, 1983, p. 419) 'The monopsonist position of most parastatals means that there is no competitive pressure to keep operations efficient, whereas their political connections make them immune to the ultimate commercial sanction of bankruptcy and dissolution.' (Schluter, 1984, p. 93)

are more able than their counterparts in the Third World to afford the 'hard-currency' bribes that are not infrequently made to government officials in these countries.)

The task of the following section is to show, for a sample of parastatal projects in Kenya and Tanzania, how the orientation of the parastatal holding companies described in the previous paragraph led to a technological 'fall-out' in which developed-country patterns with respect to scale, factor intensity and product characteristics were invariably reproduced.

The Technological 'Fall-out' from the Bureaucratic Mechanism: Case-Studies from Kenya and Tanzania

In each country a small sample of publicly-owned firms was selected for analysis. The purpose of these case-studies was not to attain any kind of statistically significant explanations; rather, the idea was to gain insights into the subtleties of the problem through a detailed study of the process by which decisions for technology are actually made in public enterprises. The selection of particular firms was made first for Tanzania from industries for which a considerable database had in most cases already been established (namely, sugar, textiles, footwear, detergents, printing ink, grain and oil milling) and the cases selected generally relate to the Third Five Year Plan period from July 1976 to June 1981.[18] For Kenya an attempt was made to select (where possible) government majority-owned firms from at least some of these industries and for a comparably recent period. These criteria, together with formidable problems of data availability and access, produced a sample of enterprises drawn from the following industries: textiles and sugar, salt manufacturing, bottling, cashew processing, ceramics, furfural and power alcohol. In both countries, data were sought from a variety of documents such as feasibility studies, board minutes, annual reports, ministerial papers and tender bids. However, the greater degree of access to these documents that was obtained in Kenya generally permitted a far more complete reconstruction of the technology decision process in the sample of firms selected from that country than those drawn from Tanzania (which should be borne in mind in interpreting the results).

Kenya

The case-studies from Kenya at once provide a quite striking confirmation of the decision-making process for technology that is envisaged by the notion of bureaucratic man and an almost equally notable disconfirmation of the implication drawn by Williams that the technological 'fall-out' from this process will tend to be random. The confirmatory evidence is contained in Table 3.4 which shows, for the sample enterprises, the predicted directness of the link between the choice of managing agents (who were generally also minority partners) and the source of technology. Table 3.5 however, suggests that when, as mostly in the cases studied here, it is developed-country collaborators that are involved, the technological 'fall-out' from this collaboration is more likely to be systematic than random. That is, the table tends to point to some tendency for the large scale of output to go together with a high degree of capital intensity on the one hand and a sophisticated type of product (which is often intended for export) on the other.

One could easily infer from Tables 3.4 and 3.5 an underlying decision-making process in which, because of the lack of concern with the technological aspects of the projects that is assumed to be exhibited by bureaucratic man (in the form of the parastatal holding companies), the 'fall-out' from the developed-country collaboration proceeded entirely unquestioned (a process in which the appropriateness of the technology to national development goals might, therefore, to this extent never be raised). But a close scrutiny of the actual negotiations reveals that this depiction of what occurred is only a part of a somewhat more complex story.

It is only a partial explanation, not so much because the parastatal holding companies were generally any more active technologically than is posited by the theory, but because of the fact that on numerous occasions *other* institutions, from various standpoints and at different stages of the projects, questioned the desirability of the technology that had been proposed. What becomes of interest, therefore, is the question of how the evident closeness of the 'fall-out' was able to be sustained in the face of the challenges to its desirability that were variously mounted. Four of the cases are particularly revealing in this regard.

Consider first the salt-refining and Nzoia sugar projects, in both of which the World Bank (whose approval is required for the Industrial Development Bank's investments of the World Bank's funds) was critical of the lack of attention paid in the appraisals of this holding company to the technological choice aspects of the projects, and in the former case drew attention specifically to what it regarded as the excessively capital-intensive nature of the equipment selected. The Industrial Development Bank's unenthusiastic response to the suggestion that more labour-intensive alternatives be considered is worth quoting in full because it reveals with particular clarity this institution's perception (and arguably, to some degree, rationalization) of the limits of technological choice, given the constraints imposed by the scale of output and the prior selection of developed country collaborators. Thus,

Table 3.4 Managing agents/minority partners and the source of technology in Kenyan parastatals

Firm	*Managing agents/minority partners*	*Main source of technology*
Rift Valley Textiles	Seditex (Germany)	Germany
Yuken Textiles	Jugotekstil (Yugoslavia)	?
Salt Manufacturers	Saltec International (Italy)	Saltec
Nyeri Bottling Plant	Coca-Cola Africa	?
Nzoia Sugar	Technisucre (France)	France
South Nyanza Sugar	Buckau Wolf (Germany)[1]	Germany/India
Kenya Furfural	Escher Wyss (Switzerland)	Escher Wyss
Kenya Cashewnuts	Oltremare (Italy)	Oltremare
Ceramic Industries	Ceramic Industries (private owners)	UK/Italy

[1] One of the turnkey contractors.

Table 3.5 Scale, technology and products: Kenya

	Scale	*Technology*	*Products*
Rift Valley Textiles	Annual capacity of 11,500,000 linear metres – 'a complete modern integrated textile mill'.	'All machinery and equipment selected are brand new, the most modern and of very high quality and reputation.'	'Very high quality' . . . 'an annual production of approximately 23 million square metres, of which about 7 million metres shall be exported.'
Yuken Textiles	'The largeness of premises and their division is a result of the conception of the whole factory and up-to-date technology as well as its flexibility.' . . . 'The biggest manufacturer of this line in Yugoslavia.'	'Highly developed productive technology . . . to compete with the European standards.'	'The technological part of the project provides a very high quality of the output (European level).' . . . 'Activities of the newly established enterprise would be concentrated during its first phase above all on the Kenyan market and on the markets of East African countries, later-on also on other markets.'
Molasses Plant	'The plant will be capable of producing 7300 tons per annum of 99.5% Power Alcohol.'	'The Installation shall conform to the latest and most modern technology and development currently available as of the date of this contract as to the Installation and the *volume* and *quality* of *products* produced.' . . . 'The opportunities to sell the surplus factory's products . . . are fully appreciated and exploited.'	
Nyeri Bottling Plant	'The plant will be equipped with a new sophisticated bottling line with ancillary equipment with a capacity for bottling 500 cases per working hour.' . . . 'The bottling equipment to be installed is very sophisticated.'		'a standard quality product.'

Kenya Furfural		'The high capital intensity of the project – due mainly to the high cost of the proposed plant.'	'Most of the end products would be mainly for export.'
Cashew Processing	'Mechanical processing in large-scale units with relatively high capital investment costs.'	'equipment considered the best in this field.' . . . 'Though mechanised, the project is still labor-intensive.' . . . 'All the machinery will be new.'	'producing high quality products.' . . . 'The project would generate significant additional foreign exchange earnings for the country.'

Source: Correspondence files of the Industrial Development Bank and the Industrial and Commercial Development Corporation.

> As IDB is involved mainly in fairly large scale industrial projects, for which technology has to be imported from the developed countries, the possibility of maximising employment (mainly unskilled labour) is severely limited by the inability of the foreign technical collaborator to adapt the technology involved to local needs. We do try to make changes in project design but our experience has so far shown that the scope to make substantial changes is limited, partly because of the technological environment to which the technical participants or sponsors are used. Some of the processes in any case have naturally to be sophisticated e.g. a salt refinery or fertilizer plant.[19]

Apparently not satisfied with this response, the officials of the World Bank offered a rejoinder to the effect that a less capital-intensive technology, even if less profitable, ought nevertheless to form part of the holding company's frame of reference. The question, as the World Bank officials saw it, was 'by how much [it was less profitable] and whether it would be worthwhile to accept the trade-off.' They added, however, that '*the project is far too advanced to raise this question at this stage*, this question should be raised in the context of the Government industrialization and employment policy and should be discussed between the Government, IDB and the promoters in the very early stages of project preparation.'[20]

In the case of the Nzoia sugar project, the World Bank's admonition (regarding the inadequate attention that IDB's appraisal had paid to the suitability of the technology selected by the French collaborators) seemed to be equally well founded in view of the fact that IDB's management 'was not able to assess the technical suitability of the proposed plant or to establish whether the price agreed on was reasonable and competitive' and had therefore, 'to make do with assurances received regarding Fives-Cail Babcock's many years of experience in supplying similar plants to various countries.'[21] But here, as in the previous case, the late stage in the negotiations at which this issue was raised seemed to make the initiation of any effective countervailing action very difficult.

It was noted in Table 3.4 that Seditex, the German managing agents of the Rift Valley textiles project, chose the equipment for this project from their home country. This firm had, however, been asked (apparently by the International Finance Corporation as a minority shareholder and lender) also to consider bids from Japanese and American machinery suppliers. Though considerably cheaper than the German quotations, these bids were nevertheless rejected by Seditex on the following grounds:

> From our view point, two vital aspects have to be considered. The first is operational costs, involving quality and efficiency; the second is the

practical and psychological conditions. By opting for the cheapest, we have no doubt that the mill's efficiency and fabric quality will be lower, that supervision, labour and maintenance costs, as well as consumption of electricity and spares will be higher. On the other hand, as European promoters, with our main office in Hamburg, it would undoubtedly be more difficult and costly to plan and coordinate with Japanese or American suppliers the multitude of details before and during the construction period. Loopholes in tight coordination and planning can have grave financial consequences.

We shall equally have considerable difficulties in coping with urgent problems that would probably require the manufacturer's assistance, once the mill is in operation. Moreover, our natural source of recruiting technicians is Europe. Such European technicians will have to become acquainted with Japanese or American equipment; this would involve more expenses and, in general, contacts would be more difficult and complicated.[22]

In the final example, that of Kenya Furfural, both the Agricultural Development Corporation and the Industrial and Commercial Development Corporation raised strenuous objections to the technological (and other) aspects of the project. The former noted that 'one or two of the intended equity holders and/or lenders whatever the case may be are also the suppliers of the plant' and sought reassurance that the plant would be purchased 'on an international tender basis'. Such reassurance did not however materialize since it was argued by the project leader that the plant and equipment, because of its specialized design, 'could only be supplied by Escher Wyss', and that, in consequence, an international tender was not possible. The ICDC objection focused on the high capital cost of the project and on an inability to obtain from the sponsor a detailed breakdown of this cost. In spite of these criticisms, other agencies of government, namely, the IDB, the Treasury and the Development Finance Company of Kenya, took a favourable view of the project, which was approved in apparently unmodified form.

For a variety of reasons, then – the fragmentation of government involvement, the attitude of the holding company, the stage at which objections were raised and the alleged hidden costs of cheaper machinery – the closeness of the technological 'fall-out' was able to be preserved in these cases. But apart perhaps from the hidden costs argument, these reasons offer little basis for supposing that the preservation had anything whatever to do with economic efficiency. Rather, it is probable that the outcome was inimical to this objective (the continuous financial losses made by all these enterprises are consistent with, though not of course proof of, this assertion).

Tanzania[23]

We have seen that the Kenyan case-studies generally conformed closely to the bureaucratic-man notion with respect to its prediction of a role for technology that is essentially derivative of the source of foreign (mostly developed-country) finance and related managerial inputs into the project. And in this derivative process there appeared to be a tendency for the technological characteristics of the project (namely, scale of output, factor intensity and type of product) to take on a particular systematic form. While some of the cases in Tanzania (mostly in sugar and textiles) also followed this general pattern, there were other cases that did not.

In these other cases, though there was still a large-scale bias, imparted in the manner described above, by the search for foreign finance, the type of finance that was obtained for these projects happened not to be closely tied to any particular source of technology, and because this permitted the technological choice aspects of the projects to assume a more central role, other factors, not allowed for in the bureaucratic-man formulation, came to be important. These other factors worked (as we shall see) through the same tendency towards interrelatedness of the various aspects of developed-country technology with which it was necessary in the Kenyan cases to supplement the original formulation of the bureaucratic mechanism. But the tendency was not, as there, merely appended to the mechanism as a derivative, but rather was set in train by an active set of preferences. That is, the basic components of the explanation of these cases remain the same, but the *manner* of their combination into an integrated approach differs from what has so far been the case.

Let us, however, first deal briefly with the case-studies in Tanzania that did follow closely along the lines of the Kenyan model. These were drawn mostly from the textile and sugar industries, which, it will be recalled (from Table 3.2a) received particular emphasis in the Third Five Year Plan. Interviews with the managers of the parastatal holding companies responsible for these industries revealed that the emphasis of the Plan on rapid import substitution had been incorporated into, and indeed seemed to dominate, the objective functions of these institutions. One manager in Texco (the holding company for the textiles industry), for example, described his primary concern as being 'to clothe the nation as fast as possible'. He stressed that if his concern was simply to maximize profits he would prefer second-hand machinery to the turnkey projects that were thought best to promote the rapid expansion of output. Similarly, what emerged from interviews with managers at the Sugar Development Corporation could be described as an objective function embodying the primary goals of meeting national excess demand as rapidly as possible subject to the constraint of avoiding losses. From this apparently overriding objective there followed the implication that particular types of products and techniques were ruled out of consideration *ab initio*. In particular, the managers stressed that the rapid expansion of the sugar industry based on jaggery and OPS sugar (the small-scale alternative to the large-scale production of refined sugar) would simply not have been feasible. In the case of the OPS alternative, they pointed out that some 200 small plants would have been required to replace the output of the five large ones. The managerial and supervisory problems associated with this alternative, not to mention the training of operatives in the OPS method and the difficulties of raising finance for small-scale techniques, meant to the managers that this method could at best be a supplement to a strategy based on large-scale factories.

In many of the large-scale projects in these industries to which the objectives described in the previous paragraph gave rise during the Third Five Year Plan period, the technology 'fall-out' was very closely tied, as in the Kenyan cases, to the sources of finance and other inputs that were secured for the (highly packaged) projects. In the case of the Musoma textile mill, for example, with a capacity of 25 million square metres of cloth, a general French contracting company was responsible for the financing and construction of the plant on a 'semi-turnkey' basis. This company identified not only the consortium of financing institutions in Europe (whose loans were effectively tied to procurement in each particular country) but also the machinery suppliers. Similarly, in the expansion of capacity of the sugar mills at Kilombero, Kagera and Mtibwa, turnkey tenders were floated in the countries to which the external finance was tied.

A recent study of the grain-milling industry by Bagachwa shows that although small-scale roller mills offer both more employment and higher profits (private as well as social) than large-scale milling technology, on no fewer than three occasions (in 1970, 1979 and 1983) the National Milling Corporation chose to expand its capacity on the basis of the latter, rather than the former, type of technology.[24] Though it is not the only explanation, the availability of external finance seems to have played an important role in this pattern of technological behaviour. For, as Bagachwa himself points out:

> Three of the British manufactured rollers had been financed by supplier's credit from Britain while another three German-made rollers

were financed by supplier's credit from West Germany. One can not simply dismiss this circumstance as a coincidence particularly in a situation where managers are desperate to secure funds for financing their projects.[25]

There were, however, other large-scale projects initiated during the Third Five Year Plan period, in which the technological 'fall-out' did not conform as closely to this derivative pattern. In these cases, in contrast, technological choice came to comprise a more central element of the decisions that had to be made for the projects. Three examples, all financed directly or indirectly (via the Tanzania Investment Bank) by the World Bank, may be cited to illustrate this point and to demonstrate, in particular, how other factors impart an independent upward bias to the capital intensity.

The most important of these factors is product choice, the bias from which, as the ill-fated Morogoro Shoe Company shows, can be imparted through the alleged need to produce exports to 'international standards'. The choice of product for this project was dictated almost entirely by its orientation to exports, as is made quite clear in the Bank's Appraisal Report. Thus, 'One fairly modern shoe factory persistently produces leather and canvas shoes for the internal market. The output is of acceptable quality within Tanzania, but does not meet international standards. The new shoe factory . . . would produce about 4 million pairs of shoes to international standards, primarily for export, but it is expected that a small part of the production will be sold internally.'[26] The same report also makes explicit the link between production for export and the nature of the technology that is required. To quote again from this document, 'During the detailed engineering design phase, efforts will be made to substitute labour for capital without compromising product style, quality and cost competitiveness *but the extent to which this can be done is limited in export-oriented industries.*'[27]

The case of Sabuni Industries Limited – the only parastatal in Tanzania which produces powdered laundry detergent – is a very clear example of a second main influence on product choice, namely, that which originates in historically determined tastes. The Sabuni project was conceived in the 1960s, when, following the break-up of the East African Community, supplies of the detergent Omo from Kenya were discontinued. It seemed clear to the National Development Corporation (at the time the holding company responsible for the detergent industry), that local production of a substitute was required. In the decision to replicate, or even improve, the brand that previously had been imported, what appears to have been crucial was the consideration that 'since most of the people in the country had used the Kenyan brand of detergent (Omo), which is of very high quality, it was envisaged to produce a detergent *of the same standard or even better*, in order to capture the market previously supplied by the Unilever Company in Kenya. The intention was also to produce a product which could be exported to neighbouring countries by competing effectively with other high quality brands.'[28] That the choice of this particular type of product had a decisive influence on the choice of technology for Sabuni, can be inferred from the impact it had on the tender process. The tenderers were apparently asked to supply, together with their quotations, a sample of their brands. These samples were then analysed by the research staff of the National Development Corporation, and it is surely not without significance that the Italian firm which was awarded the tender also happens to be the plant supplier to Unilever in Kenya, the manufacturers of Omo.

The final example, that of the multi-purpose oil mill at Morogoro, illustrates how engineering-man-type factors, in addition to those arising from product choice, operated at different stages of the decision process to exacerbate the capital-intensive bias of the technology that was chosen. Firstly, the tender document specified that all the equipment should be new.[29] This prohibition against second-hand equipment is not unique to the process of acquisition of oil-milling equipment; on the contrary, it is a practice which is endorsed in the bidding rules of the World Bank and by the governments of both Tanzania and Kenya.

In a second respect – which concerns once more the link between products and techniques – the specification of the tender document appears to have imparted a bias against the use of more appropriate techniques. Specifically, the document embodies requirements for buildings which have necessarily to be met by relatively sophisticated products and techniques. Thus, to quote from the document, 'Building and construction work must meet local building and hygienic requirements and standards, *which to a great extent are of British origin*.'[30] Or, to take another example, the use of locally produced roofing tiles is precluded by 'the minimum requirements for roofing materials – corrugated aluminium sheets or corrugated asbestos sheets.'[31]

If there were therefore biases in the manner in which the tender document was specified, it was also true that in the next stage – the evaluation of the tenders received – engineering-man-type factors further reduced the pressures for cost minimization in the choice of technology. At the very outset of the evaluation procedure, for example, we find that most of the many quotations were rejected on the grounds that they had not come 'from potentially competent and internationally-known companies'.[32] And in the selection from among the five contractors that did conform to this description, 'considerable emphasis' was laid on the following factors (in addition, of course, to costs): 'goodwill of the bidder in his country and abroad', 'competence and capability of the bidder to successfully complete the project', 'quality of the end products' and 'experience in export and establishment of plants in foreign countries'.[33]

One of the major technological issues that appears to have confronted the subcommittee formed by Moproco to evaluate the bids was the choice between a batch and a continuous refining process. Despite the relatively high cost of the latter this was nevertheless the method selected. With respect to the bleaching component of the refining process, the subcommittee motivated its choice with what amounts to a nearly complete statement of the engineering-man hypothesis. Thus, 'Continuous bleaching plants have gained wide acceptance as they are now feasible, but they are relatively costly owing to their expensive control instruments. . . . Continuous bleaching units are favoured by processors *because of their independence of human operation*, because they permit savings in bleaching earth and oil loss in it, finally *because they render a finished product of uniform colour*.'[34] In the opinion of the General Manager of Moproco, who was a member of the subcommittee for evaluating the tenders, the last reason – the uniformity of product quality associated with continuous refining – was perhaps most influential in the process.

The conclusion that emerges from these three cases is that active biases in selection led, through the close link between product choice and technology, to the same tendency towards reproduction of developed-country technologies that was shown to occur in the derivative process with which the other case-studies were associated. An essential difference, however, lies in the implications for policy that seeks to alter this outcome. In the one case it is the active biases that need to be tackled directly and in the other it is the *prior* forces that give rise to the derivation of developed-country technological characteristics. More on this important policy issue will be contained in the concluding section.

The Discretion of Parastatal Managers in Kenya and Tanzania

The applied section of this chapter has so far described the common origins of the bureaucratic mechanism in Kenya and Tanzania and the nature of the technological 'fall-out' from this mechanism in case-studies drawn from the manufacturing sectors of the two countries. The analysis has been concerned to discover how in these cases the chosen technologies acquired characteristics that rendered them inconsistent with many of the most important development goals (as stated in the respective Five Year Plans).

Following the conceptual framework, it remains to consider briefly the nature of the relationships between the parastatal holding companies and government agencies that allowed these outcomes to occur.

Table 3.6 contains a summary of these relationships, which, when juxtaposed, reveal (as with much else in this chapter) a remarkable similarity between the two countries. In both it appears that the failure of the elaborate control system has invested parastatal managers with a near-complete degree of autonomy and in both the weaknesses of the system derive from the same fundamental sources.

On the one hand, neither government has been able and willing to specify the trade-offs between the multiple and conflicting goals of development that, as we have seen, are required for the formulation of a national objective function, and without which the agents of government cannot effectively be controlled. One manifestation of this inability is in the lack of clear criteria for project appraisal on the part of the holding companies in both countries (and perhaps also in the orientation to private, rather than social, profitability that is generally exhibited in the appraisals). On the other hand, even if there were no such problem of formulating a reasonably unambiguous national objective function, the efficacy of the control system in Kenya and Tanzania would still be severely constrained by managerial and administrative resources that seemed to be quite inadequate to the burgeoning tasks that were imposed on them during the 1970s.[35] Indeed, following Killick, it is plausible to argue that such a constraint is necessarily entailed in a strategy that makes major demands on highly limited skilled manpower.[36]

CONCLUSIONS

Though there are undoubtedly many important respects in which the divergent development strategies adopted by Kenya and Tanzania have produced correspondingly different outcomes with regard to the factors that are thought to have determined the technological behaviour of public enterprises in the manufacturing sectors of these countries, this chapter has stressed instead the striking similarities between them. It was suggested not only that the two groups of enterprises adopted the same predominant policy goal, but also that the relevant aspects of the policy environment (both domestic and foreign) in which this goal was pursued were very similar. Moreover, the achievement of this common goal of rapid and sizeable output-increases in key sectors – given the macro environment that prevailed in the two countries; the similar nature of the constraints to

Table 3.6 Managerial discretion and the failings of the parastatal control system

	Tanzania	*Kenya*
Overall characterizations of the system	'The elaborate internal and external control system over parastatal activities, which exist on paper are in reality very weak.' (Ministry of Planning, 1982, p. 42)	'The President identified ineffective Government supervision as one of the factors contributing to the unsatisfactory performance of parastatals.' (Ndegwa, 1979, p. 10)
Managerial discretion	'It [the party] has remained an institution capable of directing the system only on the most general plane.' (Loxley and Saul, 1975, p. 62)	'The weakness of the party system means that the bureaucratic elites are left *on their own* to determine the major developmental policies.' (Oyugi, 1982, p. 70)
	'There was no clear policy especially as regards science and technology. At enterprise level, each corporation remained *free to choose* its own techniques.' (Mihyo, undated (a), p. 3)	'Functionally, parastatals do enjoy a very large degree of *autonomy* with respect to investment decisions.' (Oyugi, 1982, p. 71)
	'Parastatals are quite *autonomous* institutions. Control by their parent ministries is often slight.' (Clark, 1978, p. 185)	'The parent ministries tend to leave the organizations pretty much *on their own*.' (Oyugi, 1982, p. 67)
	'There is still a large question mark over the ability of boards effectively to control foreign management.' (Loxley and Saul, 1975, pp. 69–70)	'In many parastatals, the boards of management have failed to discharge the duties expected of them.' (Oyugi, 1982, p. 72)
	'In most of the subsidiaries of the National Development Corporation, the management has taken over the policy decision-making function.' (Mihyo, undated (b), p. 16)	'Most of the parastatals have management contracts with foreign based firms or locally based foreign owned management consultancy firms. They have a lot of influence on the operational policies of the organizations.' (Oyugi, 1982, p. 69)

Investment criteria and accountability	'in the absence of a clearly defined strategy for the socialist transformation of Tanzania there have been *no rational criteria* by which public sector investment and other spending decisions could be *assessed*.' (Loxley and Saul, 1975, pp. 65–6)	'At the moment it is difficult, in the case of many parastatals, to say who is ultimately *accountable* to the Board, the Government and the public for the proper management of their affairs.' (Ndegwa, 1979, p. 9) 'because of the confusion which exists, it is becoming increasingly difficult to say where *responsibility* for inefficiency or waste lies.' (Ndegwa, 1979, p. 3)
	'both the proposing parastatals and the central planning authorities *appear to lack clear criteria* on which to assess the alternative projects which they develop and evaluate.' (Perkins, 1983, p. 236)	'The *criteria* for accepting or rejecting projects are *vague* and have not included rigorous economic analysis.' (IBRD, 1983, p. 104)
The limited capacity for control	'The central planning ministry does not have the manpower to effectively scrutinize parastatal operations.' (Clark, 1978, p. 186) 'The country's already weak planning capacity . . .' (IBRD, 1978, p. 10)	'While government financial involvement has been expanding, the machinery to control and examine proposed investments in parastatals is inadequate.' (IBRD, 1983, p. 42)
Orientation to social vs. private profitability	'Most parastatals have used commercial profitability – discounted cash flow – as their investment yardstick. . . . It is however, apparent that . . . the use of DCF will almost inevitably give results which have little meaning in social terms.' (Loxley and Saul, 1975, p. 67)	'Firms with Government or parastatal involvement find that such agencies behave like any other investor, concerned with the protection and profitability of their portfolio.' (IBRD, 1983, p. 416)

achieving the objective on the basis of small-scale production; the tendency for developed-country offers of finance to be dominant in both and for these offers to be associated with the reproduction of technologies from these same countries (either merely derivatively when the tie to finance was direct or through an active set of biases when the tie was much less close); and given a similar weak government capacity for control, meant that the resulting technologies were in most respects inconsistent with major goals of development. In particular, the rapid and substantial increases in manufacturing output that were achieved during most of the 1970s in Kenya and Tanzania on the basis of large-scale, capital-intensive and highly packaged techniques appear to have had adverse consequences for employment, economic efficiency, income distribution, regional policy, products suitable for meeting basic needs, learning effects and the balance of payments (since the exports that were intended often failed to materialize from these projects while their import requirements were extremely heavy).[37]

It could be argued that this outcome reflects the fact that national goals other than output maximization were not in fact accorded much weight by either government. While this interpretation may be useful in explaining part of what occurred, it requires the assumption that an explicit (or implicit) trade-off between goals was actually made by policy-makers. It may well be more realistic, however, to posit an alternative decision-making process in terms of which policy-makers were essentially unaware of the conflicts (and especially the extent of the conflicts) with other goals that output maximization would entail. In relation to Tanzania, Hyden, for example, has suggested that 'policy makers often decide on matters without first having obtained full and detailed knowledge of the possible consequences of their decisions. They start "running" and take the consequences as they occur.'[38]

There are two broad implications for policy that follow from this interpretation of what occurred. One is that a less rapid and more moderate target increase in output would itself tend to make more feasible a strategy of relying more heavily on small-scale, labour-intensive techniques, since the managerial and organizational constraints on such a strategy would then become binding less quickly (and to this extent dependence on foreign sources of finance would also be reduced). The second broad policy option arises from the fact that even if this goal *cannot* be altered (for instance because of political factors), one can still seek to alter aspects of the environment in which the goal is pursued, so as to render its achievement more consistent with a different and more appropriate set of techniques than those with which it has been associated in the past.

The components of such an alteration, in turn, can be grouped into two broad categories, comprising policies to enhance the role of small-scale, labour-intensive projects on the one hand, and those seeking to alter the choices that are made by given large-scale projects on the other. With respect to the former, since the goal of rapid and large increases in output is assumed to remain essentially intact, and that there is therefore from this point of view no diminution in the existing constraints on an enlarged role for the small-scale sector, it is these constraints themselves that will need to be surmounted. For this purpose, it is not sufficient merely to advocate intensive training schemes that will produce additional entrepreneurs, for what will clearly be required in addition are policies to strengthen the regional, district and village organizations within which the expanded numbers of small-scale projects will have to operate.

The need for simultaneous efforts to improve the technological component of large-scale projects derives from the fact that one cannot realistically expect any rapid and substantial progress from even effective policies that are indicated in the previous paragraph. A parallel set of policies for the large-scale sector ought to comprise, firstly, efforts to restore the link between efficiency and profitability at the enterprise level. In Tanzania this would require reform of the price-control system; in Kenya it would mean, among other things, putting an end to the practice of granting privileges to public enterprises that are in financial difficulties. Particularly if they are combined with more of an emphasis on profitability in the incentive (and accountability) systems, and a general move towards a more competitive environment (as is now occurring to some degree in Kenya), these policies ought to reduce the existing degree of slackness in the decision-making procedures of the holding companies, with respect to both their search behaviour and appraisal methods. In this way, the behaviour of managers will be subject to a discipline that will make it considerably more difficult for them to ignore the technological choice aspects of projects.

Secondly, a set of policies will be required to address the problems of the finance bias that tend to be especially pronounced when there is a close 'fall-out' from developed-country sources of finance. One way to overcome this bias is to promote a much more active search for developing-country sources (e.g. Indian and Chinese), with which more appropriate techniques are likely to be associated, and a similar search for more multilateral sources, offering a wider range of technological options. But in the latter case, unless the active biases that were found to reside in the selection mechanism can be overcome, little will be gained from the wider choice that this type of finance potentially affords. Policies to correct these selection biases will have to pay particular attention to the various influences that govern product choice. For one thing, any rationale of the choice of sophisticated products that is based on the need for exports (especially to developed countries) ought to be very carefully evaluated. In many of the sample projects in Kenya and Tanzania this argument was advanced in feasibility studies (partly to rationalize particular product and process choices) but in most cases the intended exports failed to materialize and the output ended up being sold to local consumers, for most of whom it was highly inappropriate. Where the source of the product-choice bias originates instead in historically conditioned tastes for advanced-country imports, countervailing pressures – exercised, for example, through promotional campaigns for local goods – are required. While product policies of this kind may go some way towards combating engineering-man-type biases, other areas of this problem, such as the prohibition of used equipment, need to be tackled directly. Finally, because there are limits to the extent to which the sources of finance can be altered along the lines suggested above, dependence on developed-country technologies that are closely tied to sources of finance in these countries will clearly remain an important problem for both Kenya and Tanzania. In these cases, independent technological advice that is taken *early* in the appraisal of the project (from say UNIDO), and the use of some form of social rather than private mode of project appraisal, can do much to reveal grossly inappropriate choices and to increase the bargaining strength of the government *vis-à-vis* foreign collaborators.

Finally, the political economy aspects of these policy proposals need to be considered. In particular, our concern is to indicate that each of the different sets of prescribed policies will be associated with a different – albeit to some degree overlapping – profile of gaining and losing interests. And in so far as the political significance of these interests varies within and between the two countries, so too will the feasibility of the different policies.

Table 3.7 represents a crude matrix of likely gainers and losers for the three policy areas identified earlier, namely, reform of the macro environment, enlargement of the role of small-scale, labour-intensive projects and improved choices on the part of given large-scale projects.

Reform of the macro environment would have the general effect of shifting the distribution of gains from inefficient enterprises towards those that are more efficient and to consumers and firms who were previously reliant on the expensive output of the inefficient firms. Some evidence for Kenya indicates that the magnitude of these potential gains may be very substantial if protectionist policies are eliminated or reduced, since 'It is possible to document high local prices and/or low or deteriorating quality

Table 3.7 A matrix of gaining and losing interests

Interest groups/Policy	*Major interest groups that stand to lose*	*Major interest groups that stand to gain*
Reform of the macro environment	For Tanzania, those dependent on inefficient firms (public and private) that are sheltered by the price-control system	Efficient firms, firms purchasing inputs from currently inefficient enterprises, consumers in general
	For Kenya, those currently gaining most from government concessions to inefficient public enterprises	Competing (mostly private) firms, firms purchasing inputs from currently inefficient public enterprises, consumers in general
Altered composition of output in favour of small-scale enterprises (particularly those in rural areas)	Urban elites (managers, highly-paid workers, bureaucrats, consumers), foreign interests that are allied to urban elites (investors, machinery sellers, managers, skilled workers, etc.)	Rural interests (small-scale firms, the unemployed/underemployed), consumers of products made by small-scale firms (i.e. mostly the poor)
Altered technological choices made by large-scale public enterprises	Developed country interests (aid donors, managers, machinery sellers, workers), consumers of products of developed country technologies, and local interests that most benefit from the presence of developed country interests (e.g. local politicians serving on the boards of joint ventures between foreign and state capital)	Developing-country interests (aid donors, financiers, sellers of machinery), sellers of used equipment, the unemployed, consumers of products produced by the alternative technologies

for virtually the complete range of products for which there are import restrictions'.[39] The ban on competing imports of lead pencils, for example, amounts to an implicit tariff of 208 per cent, while in the case of ballpoint pens the domestically produced item sells for KSh1/75 as against a comparable import price of half a shilling.[40]

Whereas reform of the macro-economic environment indirectly promotes small-scale projects to the extent that these are relatively efficient, the second policy area shown in the table seeks directly to alter the composition of output in favour of these units, and particularly if it is pursued in a context of rural industrialization poses a challenge principally to urban and allied foreign interests on behalf of rural groups. In this case, as with macro reform, gains accrue to consumers, but because the gains derive here from the altered characteristics of the products produced on a small scale, rather than from a fall in the price of *existing* goods, they tend to be concentrated on those consumers with low incomes.

Unlike the policy prescription considered in the previous paragraph which challenges the interests associated with large scale *per se*, the final policy alternative impinges only on the particular *form* that these interests currently take, namely, those based on the predominance of developed-country finance and techniques. As such, policies to alter the techniques chosen by large-scale public enterprises threaten a rather narrower set of interests than those that seek to undermine the enterprises themselves.

Assessing the relative political strengths of the gaining and losing interests shown in Table 3.7 – and hence the feasibility of different policies within each country – is of course far from easy, not least because of the substantial disagreement among students of political economy issues in Kenya and Tanzania.[41] Nor, despite the socialistic orientation of Tanzania's leaders, can one cogently argue that the challenges to existing interests that are required will necessarily be easier to effect in that country than in Kenya. For, 'What is clear in Tanzania is that the conversion to socialism occurred essentially at the top, indeed at the very top, and has not yet permeated the whole society. As a result there remain many anti-socialist forces in the society. . . . The importance of the urban elite in derailing attempts at radical change has been emphasized by many writers.'[42] But one should not infer from this that *all* the areas of policy intervention will be no more feasible politically in Tanzania than in Kenya. It may well be the case, for example, that because of her greater degree of political independence, Tanzania is better placed to implement the switch from a predominantly developed country to alternative (notably developing-country) sources of finance and technology, as discussed in Chapter 5.

Notes

1. Clark (1978).
2. Perkins (1980).
3. Gillis (1980), p. 3.
4. Stewart (1977).
5. Okumu (1979), pp. 244–5.
6. Ibid.
7. Stewart (1977), p. 7.
8. Ibid., p. 17.
9. Ibid., p. 12.
10. Vernon (1981), pp. 10–11.
11. Raiffa (1981), p. 57.
12. Ibid., p. 62.
13. Vernon (1981), p. 12.
14. Aharoni (1981).
15. IBRD (1977a), Appendix V.
16. Clark (1978), p. 182.
17. Ibid., p. 213.
18. For details see James (1985).
19. Correspondence files of the project made available by the Industrial Development Bank.
20. Ibid.
21. Ibid.
22. Correspondence files for the project made available by the Industrial and Commercial Development Corporation (ICDC).
23. This section draws on the working paper in which the results of the Tanzania survey were first reported. See James (1983).
24. Bagachwa (1992).
25. Ibid., p. 105. Another recent study of four manufacturing industries in Tanzania comes to a very similar conclusion about the nature and extent of the influence exerted by external project finance on the technological behaviour of state-owned enterprises. In particular, the study concludes that the

    > choice of production technologies has been determined by the source of investment finance. Contrary to the intentions to employ simple technologies in which local procurement of components could be promoted, the technologies selected have tended to be sophisticated reflecting the technologies available in the financing country at that time. . . . the choice of technology has not considered indigenous alternatives. See Wangwe (1992), p. 280.

26. IBRD (1977b), p. ii.
27. Ibid., p. 29 (emphasis added).
28. From a document prepared by the National Chemical Industries at the request of the author (emphasis added).
29. Industrial Studies and Development Centre (1975a), p. 14.
30. Ibid., p. 13.
31. Ibid.
32. Industrial Studies and Development Centre (1975b), p. 2.

33. Ibid., p. 3.
34. Ibid., p. 14.
35. For more detail see James (1985).
36. Killick (1978).
37. Part of the reason for the failure of exports to materialize in Kenya had to do with the fact that the economic environment, in the form of overvalued exchange rates and a highly protected domestic market, made exporting relatively unprofitable. That is to say, the projects were set in the overall context of import-substituting industrialization rather than in the framework of East Asian style 'export substitution'. For a full discussion see IBRD (1983).
38. Hyden (1979), p. 97.
39. IBRD (1983), p. 415.
40. Ibid.
41. Some of the debate on Kenya is contained in Fransman (1982) and Swainson (1980). For Tanzania see Coulson (1982).
42. Clark (1978), pp. 204–5.

References

Y. Aharoni (1981). 'Managerial Discretion', in R. Vernon and Y. Aharoni (eds), *State-Owned Enterprise in the Western Economies*, St. Martin's Press.

M.S. Bagachwa (1992). 'Choice of Technology in Small and Large Firms: Grain Milling in Tanzania', *World Development*, vol. 20(1).

A.H. Barclay (1977). *The Mumias Sugar Project: A Study of Rural Development in Western Kenya*, PhD dissertation, Columbia University.

W.E. Clark (1978). *Socialist Development and Public Investment in Tanzania*, University of Toronto Press.

A. Coulson (1982). *Tanzania: A Political Economy*, Oxford University Press.

M. Fransman (ed.) (1982). *Industry and Accumulation in Africa*, Heinemann.

M. Gillis (1980). 'The Role of State Enterprises in Economic Development', Harvard Institute for International Development, Discussion Paper no. 83.

P. Hopcraft (1979). 'Industrialization, Balance of Payments and Trade Policy in Kenya: Effects of Protectionism and Government Intervention on Prices, Exports and Income Distribution', mimeo, Institute for Development Studies, University of Nairobi.

P. Hopcraft and J. Oguttu (1982). 'Parastatal Development Agencies and their Relationship With the Private Sector', Institute for Development Studies, University of Nairobi, Occasional Paper no. 39.

G. Hyden (1979). 'Administration and Public Policy', in J. Barkan and J. Okumu (eds), *Politics and Public Policy in Kenya and Tanzania*, Praeger.

IBRD (1977a). *Tanzania: Basic Industry Report.*

IBRD (1977b). *Tanzania: Appraisal of the Morogoro Industrial Complex.*

IBRD (1978). *Tanzania Second Cashewnut Development Project, Staff Appraisal Report.*

IBRD (1983). *Kenya: Growth and Structural Change.*

G. Ikiara (1991). 'The Role of Government Institutions in Kenya's Industrialization', in P. Coughlin and G. Ikiara (eds), *Kenya's Industrialization Dilemma*, Heinemann.

Industrial Studies and Development Centre (1974). *Tender Document for a Turnkey Project for a Multi-Purpose Oil Mill Company*, Dar-es-Salaam.

Industrial Studies and Development Centre (1975b). *Evaluation of Tender Bids for a Multipurpose Oilseed Processing Plant for Morogoro*, Dar-es-Salaam.

J. James (1983). 'Bureaucratic, Engineering and Economic Men: Decision-Making for Technology in Tanzania's State-Owned Enterprises', ILO, World Employment Programme Research, Technology and Employment Programme, Working Paper no. 125.

J. James (1985). 'Bureaucratic, Engineering and Economic Men: Decision-Making for Technology in Tanzania's State-Owned Enterprises', in S. Lall and F. Stewart (eds) *Theory and Reality in Development: Essays in Honour of Paul Streeten*, Macmillan.

T. Killick (1978). *Development Economics in Action: A Study of Economic Policies in Ghana*, Heinemann.

J. Loxley and J. Saul (1975). 'Multinationals, Workers and the Parastatals in Tanzania', *Review of African Political Economy*, no. 2.

P. Mihyo, 'Bargaining for Technology in Tanzania's Public Enterprises: Some Policy Issues', undated mimeo (a), University of Dar-es-Salaam.

P. Mihyo, 'The Legal Environment and the Performance of Public Enterprises in Tanzania', undated mimeo (b), University of Dar-es-Salaam.

Ministry of Planning and Economic Affairs (1982). *Structural Adjustment Programme for Tanzania*, Dar-es-Salaam.

R. Mukandala (1988). *The Political Economy of Parastatal Enterprises in Tanzania and Botswana*, PhD thesis, University of California, Berkeley.

P. Ndegwa (chairman) (1979). *Review of Statutory Boards*, Report and Recommendations of the Committee appointed by His Excellency the President, Nairobi.

P. Ndegwa (1982). *Working Party on Government Expenditures*, Report and Recommendations of the Working Party appointed by His Excellency the President, Nairobi.

P. O'Brien (1982). 'Relations Between Transnational Corporations and Public Enterprises in Developing Countries, with Particular Reference to Technological Development: A Preliminary Analysis', mimeo, UNIDO, Vienna.

J. Okumu (1979). 'Foreign Relations: Dilemmas of Independence and Development', in J. Barkan and J. Okumu (eds) *Politics and Public Policy in Kenya and Tanzania*, Praeger.

W. Oyugi (1982). 'Government and Public Enterprises: Some Observations on Kenya', Institute for Development Studies, University of Nairobi, Occasional Paper no. 39.

F. Perkins (1980). 'Technological Choice, Industrialisation and Development: The Case of Tanzania', DPhil. thesis, University of Sussex.

F. Perkins (1983). 'Technology Choice, Industrialisation and Development: The Case of Tanzania', *Journal of Development Studies*, vol. 19.

H. Raiffa (1981). 'Decision Making in the State Owned Enterprise', in R. Vernon and Y. Aharoni (eds), *State-Owned Enterprise in the Western Economies*, St. Martin's Press.

M. Roemer, G. Tidrick, and D. Williams (1976). 'The Range of Strategic Choice in Tanzanian Industry', *Journal of Development Economics*, vol. 3.

M. Schluter (1984). 'An Institutional Framework for Developing Kenya's Food and Beverage Exports', Institute for Development Studies, University of Nairobi, Working Paper no. 407.

F. Stewart (1977). *Technology and Underdevelopment*, Macmillan.

N. Swainson (1980). *The Development of Corporate Capitalism in Kenya 1918–77*, University of California Press.

R. Vernon (1981). 'Introduction', in R. Vernon and Y. Aharoni (eds) *State-Owned Enterprise in the Western Economies*, St. Martin's Press.

S. Wangwe (1992). 'Building Indigenous Technological Capacity: A Study of Selected Industries in Tanzania', in F. Stewart, S. Lall and S. Wangwe (eds), *Alternative Development Strategies in SubSaharan Africa*, Macmillan.

D. Williams (1975). 'National Planning and the Choice of Technology: the Case of Textiles in Tanzania', Economic Research Bureau Paper no. 75.12, Dar-es-Salaam.

4 Bureaucratic versus Economic Man: A Note on Power Alcohol Production in Kenya and Zimbabwe

The two previous chapters have taken an explicitly comparative form: Chapter 2 was concerned to explain the varying technological behaviour of public enterprises in the textile sector of a single African country (Tanzania), while Chapter 3 had as its primary goal the explanation of similar patterns of technological behaviour in two countries whose ideological differences might instead have been expected to produce dissimilar patterns.

As its title suggests, this chapter also is based on a comparative analysis, but unlike the two previous chapters the argument here is not that the objects of the comparison are equally amenable to an explanation in terms of the bureaucratic choice of techniques. The comparison is rather with that type of decision-making itself. In particular, by comparing the Kenyan experience in power alcohol production with the strikingly different outcome that can be observed in Zimbabwe, we attempt to provide an answer to the following question: How much of a difference would it make if, instead of trying to maximize the inputs of capital and foreign exchange (in accordance with the postulated behaviour of 'bureaucratic man'), circumstances were such that the firm sought rather to *minimize* these magnitudes (behaviour which, as we shall see, is far more consistent with traditional 'economic man')?

THE KENYAN CASE

As described in a number of detailed case studies on the topic,[1] the first two power alcohol projects in Kenya were approved by the government in 1977; one project had been proposed by the Mehta Group and was to be located in Muhoroni, while the other had been submitted by the Madhvani Group and was to be located in Kisumu. In both projects, as Table 4.1 indicates, the majority of the equity investment was held by the Kenya government and both were financed primarily by foreign loans of one kind or another. With respect to the Mehta Group project, this type of finance

Table 4.1 The financial structure of two power alcohol projects in Kenya

		US$ million	*(per cent)*
Madhvani			
Equity:	Kenya Government	12.0	(19.9)
	Madhvani Group	3.0	(5.0)
	Chemap Group	5.0	(8.3)
Loans:	Ex Switzerland	24.7	(40.8)
	Ex Austria	14.8	(24.4)
	Commercial	1.0	(1.6)
Total:		60.5	(100.0)
Mehta			
Equity:	Kenya Government	5.3	(26.3)
	East African Sugar Industries	0.7	(3.3)
	Mehta Group and Associates	0.7	(3.3)
Suppliers' credit:		10.6	(52.8)
Government loan:		2.5	(12.3)
Bank facilities:		0.4	(2.0)
Total:		20.2	(100.0)

Source: UNCTAD (1985), Table II.3

seems to have had much the same close and unfavourable impact on technology as was described in the two previous chapters as being characteristic of the bureaucratic choice of techniques. As in many of those examples, there was certainly a choice of technology for the Mehta plant: a choice, in particular, between a batch fermenter and a continuous fermenter. The latter method was favoured, even though, in the words of one of the detailed case-studies of the project, 'batch systems are known to be easy to build, flexible to operate and relatively stabilized technologies with a reliable track record . . . the technique chosen . . . was quite out of keeping with the factors of production available in the country'.[2]

What is also somewhat puzzling about the Mehta project is its related neglect of known and more appropriate technologies from other developing countries. In particular:

> both India and Brazil have well established ethanol industries which have evolved under constraints similar to those prevailing in Kenya at the time. Hence, in terms of cost and skill mix, the technology found in those countries would be more appropriate for Kenya than those which have evolved in developed countries. In addition, the Mehta group has

strong ties with India, and most of the technical staff running the plant at Muhoroni are expatriate Indians. Given these factors, it is surprising that the technology chosen was not Indian or Brazilian.[3]

Again as in many of the examples from the two previous chapters, however, the technology associated with this project was less the result of a calculated choice between available alternatives, than it was the direct derivative of a particular source of external finance. Thus, to quote the conclusion reached by an UNCTAD study of the power alcohol industry in Kenya, 'The choice of technology [in the Mehta project] was strongly influenced by the availability of finance from Austria'.[4] The same Austrian firm that prepared the feasibility study, for example, also supplied the machinery and contracting services, which, in their turn, were financed mainly by a supplier's credit from several Austrian financial institutions (see Table 4.1).

So close a link to external finance was not however apparent in the second of the two power alcohol projects described in Table 4.1. For, although the association with a Swiss-based engineering consultancy firm was linked to Swiss financial participation in the Madhvani project, the technology itself was purchased instead from an independent supplier (rather than from the consultant engineering firm as in the Mehta case). Any tendency to limit the capital and foreign exchange cost of the plant and equipment that this independence might have entailed, however, was offset by the terms of the management agreement, which provided for a fee that was equal to a fixed percentage of the total capital costs of the project. This provision led, predictably, to a series of attempts to 'make the capital investment as high as possible'.[5] The technical specifications, for example, contained a number of rather blatant attempts to increase the capital costs of the project, as when expensive stainless steel parts were included in place of the cheaper materials that were customarily used.[6] Investment costs were inflated too by the design of the plant to a capacity that vastly exceeded what the domestic market could bear.[7]

In one way or another, therefore, financial factors seem to have been largely responsible for the excessive capital costs of the two power alcohol projects in Kenya.[8] It was not, as we have seen, that there were no technological alternatives to those that were ultimately adopted in those projects. Nor was it the case that no alternative projects had been submitted to the Kenya government. On the contrary, even by 1976, a much cheaper proposal had already been submitted by Tate and Lyle to produce 32,000 litres of alcohol per day at a total investment cost of 2.5 million US dollars. In comparison with this proposal (which, again surprisingly, was rejected by the government), the cost of plant and machinery for the Mehta project amounted to six times as much for only twice the capacity!

The technological problems of the two Kenyan power alcohol projects were not however confined only to the thus inflated capital and foreign exchange costs; for, what had also been overlooked in these projects was an important opportunity to further develop emerging local technological capabilities. In particular:

> there was no effort in either the Mehta or the Madhvani project to encourage local participation in the projects in order to minimize foreign exchange expenditure. It has been argued that the country had no capability to fabricate any of the power alcohol equipment or its components and therefore a policy to encourage local contribution would either have been ineffective or have made the plants even more expensive. However, it has been found that some technical capacity for capital goods production does exist in Kenya, so it would have been possible for the Government to insist on some degree of local content and training to enhance the country's indigenous technical capability and possibly reduce the cost.[9]

Part of the problem, it seems, was that the government at the time had no clear policy towards technology acquisition, or indeed towards any other important aspect of the role played by technology in the development process. As in most other African countries, decisions were then being made which contained significant implications for technology, but those implications were rarely considered in advance or for that matter *ex-post* either.[10] Since then and partly on account of the disappointing experience with the power alcohol projects themselves, a number of specific changes have been made to the procedures governing investment projects in the public sector (indeed, according to Juma, most of the lessons from those projects are already reflected in existing policy.)[11] However, it remains to be seen whether and to what extent the change in government policy will influence the technological outcomes of future projects in the industrial sector. For, particularly when they embody a sizeable political component, public sector investment projects often seem to produce results very different from what policy makers are genuinely trying to achieve. This problem has been mentioned in earlier chapters, where several examples were cited, but it receives particular emphasis later, when we turn to political economy in Chapters 7 and 8.

THE ZIMBABWE EXPERIENCE

Zimbabwe's early experience with power alcohol production could hardly offer a more striking contrast with the cases that have just been described. A substantial part of this difference, it seems, can be ascribed to the fact that in Zimbabwe the institution concerned was bent on minimizing, rather than maximizing, the investment and foreign exchange costs of the project. Compared to the Kenyan case, what was distinctive about the institutional environment that gave rise to this objective, was not so much that the first power alcohol project was undertaken by a private rather than a publicly owned firm. It was rather that the enterprise concerned had no access to foreign exchange other than through the government, and such amounts of foreign currency as were made available had to be recovered within a six-month period.

This requirement, in turn, exerted a powerful influence on the technology that was selected at Triangle (a major sugar estate in Zimbabwe), as the following description of the decision-making process quite clearly suggests. Thus to begin with:

> considerable thought was given by the management of Triangle to the choice of technique. Given the lack of foreign exchange and the limited availability of both skilled and unskilled labour, one of the prime requirements was that the plant be simple so that it could be constructed cheaply and operated easily. . . . While continuous fermentation was looked at, it was rejected because it was regarded as too complex at the time and involved the external purchase of . . . centrifuges at $100,000 each. . . .
>
> The source of the technology was decided after senior staff visited France, Brazil, the Federal Republic of Germany. . . .
>
> The design finally chosen was from Gebr. Herrman/Buckau Walther of the Federal Republic of Germany and consisted of simple conventional batch fermenters. The reasons for this choice were that the basic technology was simple to construct, in contrast to continuous fermentation systems, and was well-proven.[12]

It is also worth noting in this regard that the capital costs of the plant and equipment were reduced by deliberate efforts to use less-expensive grades of material, wherever this was permitted by the technical specifications. Compared to the Kenyan case, therefore, the scarcity of foreign exchange imparted a very different, and indeed, in some respects a diametrically opposite, type of influence on the technology adopted at Triangle.

For, not only did the choice of technology itself become a central rather than a neglected dimension of the project, but the direction of the influence thus exerted was towards simple and inexpensive, not complex and capital-intensive, technology. Not surprisingly, moreover, the need to minimize the foreign exchange costs of the project bore rather heavily on the choice between local and imported capital equipment, and hence also on the extent to which indigenous technological capabilities were promoted by the Triangle venture.

Encouraged as it was by the comparative simplicity of the technology and by the existence of a comparatively well-developed capital goods sector in relation to most other African countries, the import demand at Triangle was lessened by the local fabrication of a 'substantial fraction of the plant'.[13] Indeed, much of the plant was fabricated and constructed on the site of the project itself. 'This entailed setting up a school to train local people in fabricating skills such as welding mild and stainless steel. The training period was brief (2–3 weeks), and some of the less critical components in the plant were fabricated during this period.'[14] Local technological capabilities were subsequently enhanced still further, by means of a system of incentives designed 'to ensure that the plant was fabricated to the stringent requirements of the manufacturer'.[15]

In making a case for policy on the basis of this experience, one should emphasize that it is not only in privately owned firms or in African countries with relatively well-developed capital goods sectors, that a perceived scarcity of foreign exchange has been known to stimulate the acquisition of local technological capabilities. Mytelka, for example, recounts the case of Kiltex, a publicly owned textile firm in Tanzania, which, in the mid-1970s, was confronted with an acute shortage of foreign exchange.[16] The firm's engineering department responded to this situation with a number of initiatives that had the effect of substituting local for imported technologies and other related inputs. These initiatives included the production of spare parts and a series of innovations that made more effective use of, or reduced the need for, imported inputs of one kind or another. (Mytelka aptly describes these efforts as 'steps in a process of technological apprenticeship'.[17]) It seems therefore that even in an environment as generally unconducive to the acquisition of technological capabilities in the public sector as was Tanzania during the 1970s, pressure can nevertheless be brought to bear on the individual enterprise that facilitates rather than suffocates the process. To cite Mytelka again, what appears to be crucial in this regard is a perceived need on the part of the enterprise 'to reduce costs of production, whether these be costs of foreign exchange, energy, or expatriate salaries. It is this need to reduce the costs of production that thus provides a powerful stimulus to technological mastery as it inevitably requires the assimilation of imported know-how by local personnel and the replacement of imports by domestic inputs'.[18]

BUREAUCRATIC MAN VS. ECONOMIC MAN

Thus summarized, the decision-making process at Triangle has a rather close affinity with the traditional micro-economic theory of the enterprise. The search for information about, and careful comparisons between, alternative technologies, closely reflects the highly calculated type of behaviour that this theory implicitly assumes.[19] But at the same time, because the selection and subsequent use of the technology were governed by the need to minimize the capital and foreign exchange costs of the project, the outcome was consistent with what one would generally expect to follow from such behaviour, namely, an economically efficient process of production and a low ratio of capital to output. (In fact, in the mid-1980s it was possible to go so far as to state that 'due to careful management and choice of technique, the capital cost of the plant per unit of output is the lowest recorded of any plant operating in the world'.[20]) In these various respects, what occurred in Zimbabwe conforms very closely to several other successful privately owned firms in other African countries that are described in Chapter 5 (cases which also exhibit, for example, substantial technological effort and the choice of relatively simple, low-cost technology that can be readily repaired and maintained).

In each of these respects, however, there is a striking contrast between the behaviour of 'economic man' and that exhibited by the two state-owned enterprises in Kenya, which we suggest conforms instead to the so-called 'bureaucratic' choice of techniques. Just how much of a contrast these cases actually represent, is perhaps most easily seen from the juxtaposed descriptions contained in Table 4.2. The purpose of this comparison, one should note, is not at all to suggest that the one form of behaviour is in some sense rational while the other is not. The essential point is rather that the actors involved were responding to very different institutional environments, associated, among other things, with contrasting incentive systems. In thus emphasizing the influence of environmental factors on the observed behavioural outcomes, it has to be said however that the evidence on which we have drawn does not control in any rigorous way for possible differences in the firms themselves. Because it is therefore possible that these, rather than environmental, differences explain some part of the variations in technological behaviour between the firms in the

Table 4.2 Decision-making for technology: power alcohol in Kenya and Zimbabwe

	Kenya ('bureaucratic man')	*Zimbabwe ('economic man')*
Technological effort (i.e. the search for and comparison of technological alternatives)	Limited	Intensive
Reliance on the domestic capital goods sector	Non-existent	Substantial
Enterprise goals in relation to foreign finance	Maximization	Minimization
Enterprise goals in relation to capital costs	Maximization	Minimization
Relationship between the supply of technology and external project finance	Very close*	Unrelated
Nature of chosen technology	Complex, highly packaged, overbuilt and very high ratio of capital to output*	Simple, unpackaged and highly efficient, with an exceptionally low ratio of capital to output

* Refers especially to the Mehta Group project in Kenya.

two countries, the interpretation we have advanced should be treated with the appropriate degree of caution.

From several of these points of view, moreover, it appears that the power alcohol experience was by no means untypical of the then prevailing technological behaviour in Zimbabwe's industrial sector. Watanabe, for example, has suggested that the effect of international sanctions against what was then Rhodesia, had a generally favourable impact on local technological capabilities in the industrial sector, partly because of the intensive technological efforts that were thereby stimulated.[21] In particular, he suggests that by limiting the amount of foreign exchange that could be used to import technology, sanctions induced firms to identify and to solve technological bottlenecks by their own in-house efforts. For these purposes, firms tended to become highly active technological institutions, whose keen search for know-how sometimes entailed overseas visits and which, when successful, often led to an exchange of information among local industrialists.

CONCLUSIONS

The first part of this chapter, which dealt with two power alcohol projects in Kenya, described a pattern of technological behaviour distinctly reminiscent of the other Kenyan case-studies described in Chapter 3, even though these two projects were not studied from the same theoretical perspective adopted there.

The second part was concerned with a power alcohol project in Zimbabwe, where both the decision-making process and the technological outcome reflect a very different institutional environment: an environment that led most notably to a tendency to minimize rather than maximize the capital and foreign exchange costs of the project in general and of technology in particular. In the process and in common with other firms having to confront the effects of international sanctions, local technological capabilities appear to have been stimulated rather than discouraged or left unaffected.

Notes

1. UNCTAD (1985) and Juma (1991). The discussion that follows is based on these two sources.
2. UNCTAD (1985), pp. 17 and 56.
3. Ibid., p. 56.
4. Ibid.
5. Juma (1991), p. 33.
6. Or, in an even more extreme example, 'the support structure of the distillery was . . . built to withstand earthquakes measuring 6.0 on the Richter scale'. Ibid.
7. For a variety of reasons, the capital costs of the Madhvani project rose during the construction period until 1982, when the project was discontinued. At that point, almost 200 million US dollars had been spent on the project.
8. As in many other public sector investment projects in Kenya, more or less strenuous objections to the power alcohol projects were raised by certain institutions of the state. An evaluation conducted by the Industrial Survey and Promotion Centre of the Ministry of Commerce and Industry, for example, concluded that the two projects were too costly, especially the Madhvani project. As with so many of those other projects, however, the concerns were ignored and the original plans to produce power alcohol were approved.
9. UNCTAD (1985), p. 23. In particular, the metal fabrication subsector is said to be well developed in Kenya, as evidenced for example by the export of engineering equipment to other East African countries.

10. See Vitta (1990) for a recent review of technology policy weaknesses in Africa. The need for and the scope of technology policy in small developing countries are discussed in Forsyth (1990).
11. Juma (1991). He argues nevertheless for a 'viable technology policy' for Kenya which is described in terms of a 'major mobilization programme bringing the private sector, research bodies, government, and the public into new forms of institutional arrangements' (p. 50).
12. UNCTAD (1985), pp. 38 and 43.
13. From the supply side, this meant that agreement had to be obtained from the German supplier of the technology for the local fabrication of the plant. 'Apparently the company in question was the only firm to agree to these conditions and it seems that its decision was prompted by . . . the desire to be responsible for the first operating power alcohol plant in Africa and caution about being involved in overtly breaking international sanctions.' Ibid.
14. Ibid.
15. Ibid.
16. Mytelka (1985).
17. Ibid., p. 118.
18. Ibid.
19. The neo-classical model effectively assumes a highly effective degree of decision-making at all stages including search, the processing of information and the comparison of alternatives. See Leibenstein (1976).
20. UNCTAD (1985), p. 52.
21. Watanabe (1987). This article also contains a useful comparison between the patterns of industrialisation in Zimbabwe (where because of sanctions foreign exchange was scarce) and Tanzania (where, because of foreign aid, foreign exchange was relatively abundant). This comparison, he finds, bears out 'Hirschman's view that probably the most general truth about innovations is that one makes them when one must' (p. 538). What also emerges from this comparison is a distinction between technological know-how that is actively sought by the user (as in Zimbabwe), as opposed to being only passively received (as in Tanzania). In the former case local capabilities were strengthened, whereas in the latter example no such effect was apparent. Indeed, imported technologies and know-how may even have had a negative impact on local capabilities in that they discouraged 'efforts at self-reliance'.

References

D. Forsyth (1990). *Technology Policy in Small Developing Countries*, Macmillan.

C. Juma (1991). 'Investment Strategy and Technology Policy: Lessons from Power-Alcohol Development', in P. Coughlin and G. Ikiara (eds), *Kenya's Industrialization Dilemma*, Heinemann.

H. Leibenstein (1976). *Beyond Economic Man*, Harvard University Press.

L. Mytelka (1985). 'Stimulating Effective Technology Transfer: the Case of Textiles in Africa', in N. Rosenberg and C. Frischtak (eds), *International Technology Transfer: Concepts, Measures and Comparisons*, Praeger.
UNCTAD (1985). *Power Alcohol in Kenya and Zimbabwe: A Case Study in the Transfer of a Renewable Energy Technology*, TT/61.
P. Vitta (1990). 'Technology Policy in Sub-Saharan Africa: Why the Dream Remains Unfulfilled', *World Development*, vol. 18(11).
S. Watanabe (1987). 'Technological Capability and Industrialisation', *International Labour Review*, vol. 126(5).

5 Appropriate Technology in the Public Sector

The argument advanced in the earlier chapters does not entirely preclude the possibility of appropriate technology in the public sector. What was emphasized, rather, was that most offers of external finance for public sector projects in Africa emanate from developed economies and that this type of finance (in conjunction with related project inputs) is generally associated with inappropriate technology. In practice, of course, exceptions to both these generalizations occur. Finance and other project inputs do sometimes originate in developing countries, where, to a somewhat greater extent than in the industrialized market economies, conditions tend to favour the development of relatively appropriate technologies, but even developed-country sources of external finance may at times lead to appropriate forms of technology in the public sector. The purpose of this chapter is to show how a number of exceptional cases from Kenya and Tanzania can be explained in these terms and to assess the lessons for policy that such cases might suggest.

DEVELOPING COUNTRY SOURCES OF EXTERNAL FINANCE AND OTHER INPUTS

The same historical perspective that we have used before to explain the bias in favour of large-scale, capital-intensive technologies in the public sector, can also be employed to understand some of the exceptions to this general pattern of technological behaviour in Kenya and Tanzania. For, just as there is reason to expect one pattern of technical change (in products and techniques) to emerge in advanced industrial countries, so too may one expect a different and more appropriate pattern to emerge in the developing countries. And even though the former may be much the more prevalent and better-known part of the shelf of available technologies, cases where the latter appear in public sector projects in Africa cannot be ruled out entirely.

In broad terms, the pattern of technical change that might be expected to occur in developing countries corresponds to the history of the now-developed countries at a comparable stage of their development.[1] During the Industrial Revolution, for example, when a relative abundance of labour existed, innovations in the manufacturing sector tended to reflect that historical condition. In particular, evidence for the early part of the nineteenth century in Britain indicates that the direction of technical change tended to be neutral rather than labour-saving; employment and capital accumulation in manufacturing grew at the same rate of about 2.5 per cent per annum. In those European countries that were early followers of the Industrial Revolution, too, one can find evidence of strenuous efforts to adapt technologies to the labour abundance that then existed, as manifest for example, in the building of simpler and less expensive equipment and the adoption, whenever possible, of second-hand machinery.

Similar tendencies have been observed in a number of contemporary developing countries, often, it would seem, as a response to much the same need, namely, to generate and adapt techniques and products in line with prevailing factor and other resource endowments on the one hand and with levels of income on the other. Some such examples appear in studies that are concerned to compare patterns of innovation in the developing and developed countries. A detailed study by Wells, for instance, compares multinationals from the Third World with those from developed countries and comes to the conclusion that:

> No matter which comparisons one prefers, it is clear . . . that the developing country firms innovate in different ways . . . from firms originating in the industrialized countries. . . . they have found various ways to manufacture efficiently at small scale. . . . Similarly, firms from developing countries have learned to use locally available inputs to a greater extent than firms from industrialized countries. Furthermore, on occasion Third World firms produce products that are especially appropriate to developing country environments.[2]

Other examples can be gleaned from the experience of particular developing countries rather than from such comparisons.

The East Asian experience in particular most closely seems to resemble what occurred during the Industrial Revolution in Europe (not least, it seems, because of the highly competitive environment which in both cases permitted prevailing factor scarcities to induce labour-using innovations). In Korea, after the mid-1960s, for example, 'capital-stretching' innovations are said to have been abundant in textiles, electronics and plywood production.[3] Indeed, so pervasive do these forms of industrial innovation appear to have been at that time, that the capital–labour ratio in Korea actually fell for part of the period after 1964.[4] Similarly, in Taiwan after 1959 Johnston and Kilby describe the 'driving organizational abilities of management', which 'insured that every commercially feasible possibility for substituting labor for capital was exploited'.[5]

Still other examples have emerged from China and India, countries which adopted development strategies that differ not only from each other, but also (and most markedly) from the East Asian model. In the latter, the development of the Swaraj tractor as a commercially successful labour-using alternative to imported brands is one example,[6] and the often-cited improvement in small-scale open pan sulphitation (OPS) sugar technology is another.[7] In the former, imaginative use has been made of 'descaled' plants for use in rural areas, as occurs for example with cement suitable for certain rural needs.[8]

All these and further examples that could easily be cited, should not be construed as an argument that all or even most innovations in developing countries conform to this pattern. For there is a certain amount of evidence to the contrary that also needs to be taken into account: evidence, for example, that technical change in some developing countries merely replicates the developed-country pattern (India is a country often cited in this regard because of numerous observations that the production of machinery there takes a labour rather than a capital-saving form).[9] Whether, on balance, there is nevertheless a tendency for innovations to be more appropriate in the developing than the developed countries is not a question that can easily be answered on the basis of the existing somewhat fragmentary evidence.[10]

What can be said with a greater degree of certainty, however, is that such appropriate innovations as do exist in the Third World, are usually less readily transferred to state-owned enterprises in Africa and other poor countries, than technologies emanating from the developed countries. This is partly a question of finance; as noted in Chapters 1 and 3, aid and other capital-flows from developing countries constitute only a small proportion of total capital-flows to the Third World in general and to Africa in particular.[11] But it is also a question of information bias that inhibits the transfer of technology from one poor country to another. One important source of this bias is that technical change emanating from developed countries is usually heavily promoted (by, for example, machinery salesmen and consultants) in comparison with more labour-using alternatives.[12] Information biases are also due to professional predilections that favour modern and eschew relatively dated technologies.[13]

Some of the exceptional cases in Kenya and Tanzania – where appropriate technology was used in public sector industrial projects – were successful partly because they were able, in one way or another, to overcome these various obstacles.

CHINESE TECHNOLOGY IN TANZANIA

We noted in Chapter 3 that Tanzania continued to receive substantial amounts of foreign aid from developed market economies in the decade following the Arusha Declaration. Yet, some degree of donor diversification was achieved. Most notably, the period saw a fall in the share of foreign assistance provided by the United Kingdom and the United States and an increase in the share provided by socialist countries, especially China.[14] Most of the Chinese assistance was provided for the construction of the Tanzam railroad, but a number of other projects in the industrial sector were financed as well.[15]

One of these projects, the Friendship textile mill, has already been mentioned in Chapter 2, where it was shown to be amongst the most efficient firms in the entire industry. What was not mentioned, however, was that from the standpoint also of certain indigenous technological capabilities, the Friendship project appears to have performed relatively successfully. In particular 'the Chinese put great emphasis on on-the-job training, as a result of which the Chinese experts were successfully replaced within five years. It is notable that the Chinese-built textile mill is the only integrated textile mill in which foreign experts were replaced within such a short time'.[16]

A second Chinese aid project, Ubungo Farm Implements (UFI) was established at around the same time (in 1968) to produce traditional tools and implements, such as hoes and ploughs, for use in rural areas. As in the textiles case, finance was provided on the basis of an interest-free loan from the Chinese government. And there are also close similarities between the two projects with regard to the choice of technology and plant design. For one thing, the technological component of the farm implements project seemed to receive scant attention from the Tanzanians; indeed, Barker et al. can find 'no evidence' whatsoever to suggest 'that the Tanzanian side took any greater degree of interest in the choice of technology than in any other project'.[17] Moreover, for what one suspects are similar reasons (low capital costs and foreign exchange savings), the relatively simple technology favoured by the Chinese designers of the farm implements project has enabled UFI to attain a level of competitiveness in the engineering industry that is comparable to what Friendship has accomplished in textiles.[18] In yet another similarity to the latter, technological capabilities of a particular type were apparently promoted at UFI by a very rigorous on-the-job training scheme that allowed the Chinese technical personnel to be replaced within five years of the project's inception.[19]

Interestingly, the Chinese technology itself is said to have enhanced the opportunities for learning at the farm implements plant, through a subtle and somewhat unexpected set of relationships between product specialization, automation and skill requirements. In particular,

> Under the over simplifying assumptions of cost benefit analysis, the intention to produce large numbers of simple agricultural implements would lead to the installation of special purpose equipment with open or closed loop automation. And such technology exists. But the Chinese Government who built UFI decided on using general purpose machines and hand tools without any open or closed loop automation. The significance of such a decision is that the 'simpler' technology chosen demands the development of a much higher level of basic industrial skills than are required for more automated set ups. UFI thus makes a contribution to Tanzanian industrial development, in providing a place where large numbers of Tanzanians can gain these skills.[20]

Expressed in Hirschmanian terms, what this quotation raises, is the possibility of 'trait-making' behaviour[21] – behaviour, that is to say, that seeks to redress rather than to accept prevailing skills limitations during the project. If successful, such behaviour could significantly have enhanced technological capabilities in Tanzania's capital goods sector. It is not clear, however, that this potential was actually realized at UFI, where there were production problems of various kinds (such as shortages of raw materials and spare parts, water and electricity interruptions, etc.), which reduced the plant's capacity and hence its ability to impart skills to the labour force.[22] At the same time, the discrepancy between what was intended and what was actually achieved, may also reflect the fact that the original intention – on the part of the Chinese – to diversify production was later eschewed by the Tanzanians in favour of the mass production of only a few items.[23] (There are doubts too about the extent to which the experience gained at UFI has proved useful in subsequent public sector projects in the capital goods industry. Wangwe, for example, has recently suggested that learning effects were not apparent in the design of ZZK, the second large-scale state-owned producer of farm implements in Tanzania.[24] He draws particular attention to the fact that product design at ZZK was entrusted entirely to outside consultants, in spite of the extensive experience in this activity that had already been acquired at UFI.)[25]

INDIAN TECHNICAL COLLABORATION IN KENYA

Like the two foregoing examples, Pan African Paper Mills in Kenya illustrates how a connection to developing-country project inputs can give rise to appropriate technology in the public sector (though of course as noted earlier this is not an inevitable outcome).[26] Unlike those two examples, however, the connection to the Third World in the case of Pan African Paper did not arise through bilateral foreign aid. Nor did it take the form of loan finance for the purchase of machinery and equipment, which was provided in this case by North American and British sources. It took the form, rather, of a managerial and technical agreement with an Indian-based firm (Orient Paper Mills), which, together with the Kenyan government and the International Finance Corporation of the World Bank Group, sponsored the formation of the project in 1970.[27]

With what one presumes was the benefit of their experience in India, where paper production is undertaken mainly by relatively small and labour-intensive firms, Orient Paper Mills deliberately and successfully designed a pulp and paper plant that is especially well suited to Kenyan conditions. (The Indian technical collaborators were also responsible for an intensive training programme, which included among its other features, pre-operational training of Kenyans in India.) The extent and nature of these adaptations have been described – and published – with an unusual degree of technical detail; one needs to be conscious therefore of the possibility that what appears to be exceptional about this case may reflect instead the unusual degree to which its technological aspects have been documented, in comparison with other projects in the public (and for that matter also the private) sector.

All the same, it is difficult to remain unimpressed by what was achieved at Pan African Paper in the name of appropriate technology. Most immediately, perhaps, one is struck by the sheer number of respects in which the term 'appropriate' can be applied to this project. Unlike most industrial projects in Africa, for example, the pulp and paper plant in Kenya was constructed in a relatively isolated rural setting. Indeed, according to one description of the project written in 1979, 'The paper mill brought an economic transformation to a rural situation on a scale never before seen in Kenya. From a typical village of no more than 500 persons at the inception of the project, the town of Webuye in the Western Province has, over the last few years, grown into a major commercial centre with a population of more than 20,000 centred around the paper mill.'[28]

Also unlike the typical African case were a series of technological efforts undertaken specifically on behalf of appropriate technology. These took the form, partly, of technological adaptations to the natural resource environment (as when, for example, research was conducted in India to discover blends of local pine, cypress and eucalyptus fibres that would be consistent with the desired characteristics of the end product). Most forms of technological effort, however, were guided by still other dimensions of appropriateness; in particular,

a) Equipment had to be rugged, require little maintenance and be simple to operate;
b) Wherever possible, low-energy systems were selected;
c) Capital costs had to be kept down by resorting to simple design and proven conventional equipments, but without sacrificing flexibility of production capability over the entire range of products.[29]

It would be tedious to list all the respects in which the technology at Pan African Paper meets these requirements. The entries in Table 5.1, however, provide some indication of just how widely distributed over the various technical functions were the efforts made there on behalf of appropriate technology.

Interestingly, both the direction and intensity of these technological efforts bear a rather striking resemblance to what is known about the experience of a group of successful privately owned enterprises in East Africa. Barker et al., for example, have studied the technological behaviour of a sample of firms owned by domestic and East African private capital in Tanzania.[30] These enterprises seemed to the authors to have spent considerable time and effort selecting technology, which typically entailed relatively low investment costs and which was capable of being easily repaired and maintained.[31] The technological effort thus exerted by the firms concerned did not eschew outside assistance; on the contrary most of them used a combination of in-house and expatriate personnel in the process of identifying and selecting efficient machinery and equipment. What seems to matter, therefore, is not the presence or absence of foreign suppliers of knowledge but rather the extent and direction of their involvement.

Though not part of the sample studied by Barker et al., Afro-Cooling is a particularly well-known and clear example of a privately owned Tanzanian firm that has relied heavily on appropriate technology and technological mastery for its success, which, like Pan African Paper in Kenya was achieved to a large extent in export markets and which, again like the latter, involved Indian technical collaboration. The similarity between these two cases emerges even more forcibly from the following account of what occurred at Afro-Cooling:

> This firm started operations in 1979 to make radiators for the various models of automobiles on local roads. It imported a dozen technicians

Table 5.1 Appropriate technology at Pan African Paper (Kenya)

Technical function	*Elements of appropriate technology*
Pulp milling (unbleached)	'Conventional batch digesters were preferred to the more sophisticated continuous digester. The benefits of cooking by the batch system are: lower costs, less maintenance, assurance of continuity in production and flexibility to produce various paper qualities' (p. 46).
Mechanical engineering	'Direct drives with gearbox and flexible coupling were preferred, as they work even with slight errors of alignment . . . Wherever possible, drive elements were used that could be taken out without disturbing the main parts . . . The advantage is that they can take more angular and axial misalignments, which are bound to happen with limited skills in manpower (p. 49).
Electrical engineering	'An example is the salient-pole, air-cooled brushless alternator, which facilitates easy dismantling and repair . . . Additional safety interlocks are used to minimize possible damage by improper operation' (p. 50).
Instrumentation and controls	'Large instruments are preferred to small ones because of easy servicing without affecting the process. Most of the control valves are of simple design' (p. 51).
Building design	'Owing to the mild to tropical climate, building designs could be kept simple, with maximum utilization of locally available material. Reinforced concrete buildings with structural steel columns, and flat roofs with pre-cast sections were selected. This choice reduced the cost of construction' (p. 51).

Source: V.L. Moorthy (1979).

from an Indian firm with considerable experience of radiator manufacture (originally based on German technology). These experts designed a small plant with a simple two-shed layout, procured equipment from Germany and simple tools (very cheaply) from India. They trained local workers intensively, working with them on the shop-floor. Though the scale of operation was about one-tenth of international levels the use of very labor-intensive, simple techniques enabled Afro-Cooling to keep costs down and to efficiently produce an enormous variety of radiators. It was able to export its products to other countries in the region and the Middle East.[32]

DEVELOPED-COUNTRY SOURCES OF APPROPRIATE TECHNOLOGY

As we have seen, it is not invariably the case that developed-country sources of external finance lead to inappropriate technology in the public sector. One reason why exceptions to the general tendency are likely to arise is that a certain (albeit relatively small) amount of such finance originates in developed-country institutions which, for one reason or another, do facilitate the adoption of appropriate technology by public enterprises.

While appropriate technology institutions are probably the most obvious examples from this point of view (the more so because, as with the brick-making example in Chapter 2, which directly involved one such institution), certain national and international organizations also have to be considered. The International Labour Office, for example, has been an especially prominent advocate of appropriate technology over the years, in terms of both its research agenda and its policy initiatives. Indeed, one of the most successful of the projects associated with this institution was undertaken in Kenya as a follow-up to research that had been conducted by its Technology and Employment Branch,[33] research that had shown labour-intensive construction technologies to be both technically and economically variable. The Rural Access Roads Programme (RARP) in Kenya translated those results into practice through the construction of about 8000 kilometres of access roads by large numbers of village labourers (who are also used on a part-time basis to maintain the roads). The RARP in turn has been followed up by Minor Roads Programme (MRP) for which the ILO is playing the role of donor coordinator and which, like the RARP, employs mainly casual labourers who also work part-time on their own farms.

It would also be inaccurate to entirely exclude multinational corporations as agents through which appropriate technology can potentially be transferred to the public sector. Pack, for example, has argued that these firms are easily able to identify and transfer equipment among their subsidiaries, especially labour-intensive technologies that have become unprofitable in countries with relatively high wages (generally, the developed industrialized countries).[34] And at least one case in Tanzania – involving the state on the one hand and a multinational corporation on the other – seems to conform to precisely this pattern of technological behaviour.[35]

AN UNUSUAL JOINT VENTURE IN TANZANIA

Like many state-owned enterprises in the industrial sector, General Tyre (East Africa) was constructed and operated on a turnkey basis, with the American-based parent company (General Tire International) acting in this case both as supplier of the entire plant and as a minority shareholder. Unlike the vast majority of those enterprises, however, General Tire has been an efficient and profitable undertaking; indeed, it is the only example of a turnkey project in the sample of Tanzanian enterprises studied by Barker and colleagues that 'operates extremely well'.[36]

The same authors seek to explain this unusual performance in the public sector, with reference to a set of organizational factors that led the parent company to pay special attention to the profitability of its Tanzanian operations. In particular,

> The parent company not only runs the subsidiary under a management agreement, but also supplies all the raw materials, intermediate goods, spare parts and moulds. General Tire International have an interest in running the Arusha factory efficiently, as one can expect, for only when production is high can they make ongoing profits through the supply of raw materials and intermediate goods.[37]

This combination of circumstances contrasts sharply with the conditions under which most other turnkey projects in Tanzania have been supplied by foreign firms acting as minority shareholders. To quote Barker again:

> The fact that the machine merchant is willing to become a minority shareholder, does not guarantee any commitment to making a project successful in production terms. The capital provided will presumably be supplied as plant and equipment, already including an element of profit. Even if the foreign minority partner is asked to supply management along with the technology, there is no guarantee that a viable enterprise will result; other management agreements include no real incentive to attain good performance.[38]

These motivational differences between General Tyre and other turnkey projects probably help to explain divergences in their technological behaviour as well. For, whereas public enterprises in the industrial sector tended (as we have seen in Chapter 1) to use modern, highly automated technologies, the General Tyre subsidiary in Tanzania, presumably with an eye to cost reductions, imported second-hand equipment from the Dutch subsidiary of the parent corporation (an operation that seems to fit the aforementioned description of how appropriate technology may arise within the multinational corporation). And though General Tyre was not alone among public enterprises in using second-hand equipment, there were few other cases where the technology did not experience major problems and breakdowns. (The essential point as Barker and colleagues rightly emphasize, is not about 'new versus used' technology *per se*; it is rather about 'who sold what and or dumped, what upon whom, and why').[39]

APPROPRIATE TECHNOLOGY IN THE PUBLIC SECTOR: A TREE-DIAGRAMMATIC REPRESENTATION

When combined with the two previous chapters, the foregoing discussion suggests that decision-making procedures in the public sector exhibit two main features. The first is that decisions taken very early in the project cycle bear heavily on the technological outcomes of public sector projects in Africa. This influence, second, is imparted partly by the powerful influence of external finance on technology and partly also by the fact that the initial decisions themselves bear on what is decided at later stages of the project cycle. Technological outcomes of public sector investment projects therefore depend both on the probability that initial decisions will take one form rather than another, and on the subsequent extent to which these decisions influence the outcomes of later stages in the project cycle.

These ideas can perhaps best be expressed in the diagrammatic form of a 'tree' whose 'branches' and 'forks' connect decisions and outcomes during the four major stages through which public sector projects typically pass, namely, identification, design, appraisal and implementation.[40] Put somewhat differently, the idea is that there is a sequence of forks along whose various branches a project can be viewed as progressing during a particular period of time. By means of such a diagram one can illustrate not only the typical technological trajectory of public sector projects in industry, but also the particular sequence of decisions and outcomes that give rise to more appropriate forms of technology.

The first fork in Figure 5.1 reflects the view expressed in several of the previous chapters, that public sector behaviour during the project identification stage is not governed mainly by technological preferences of one

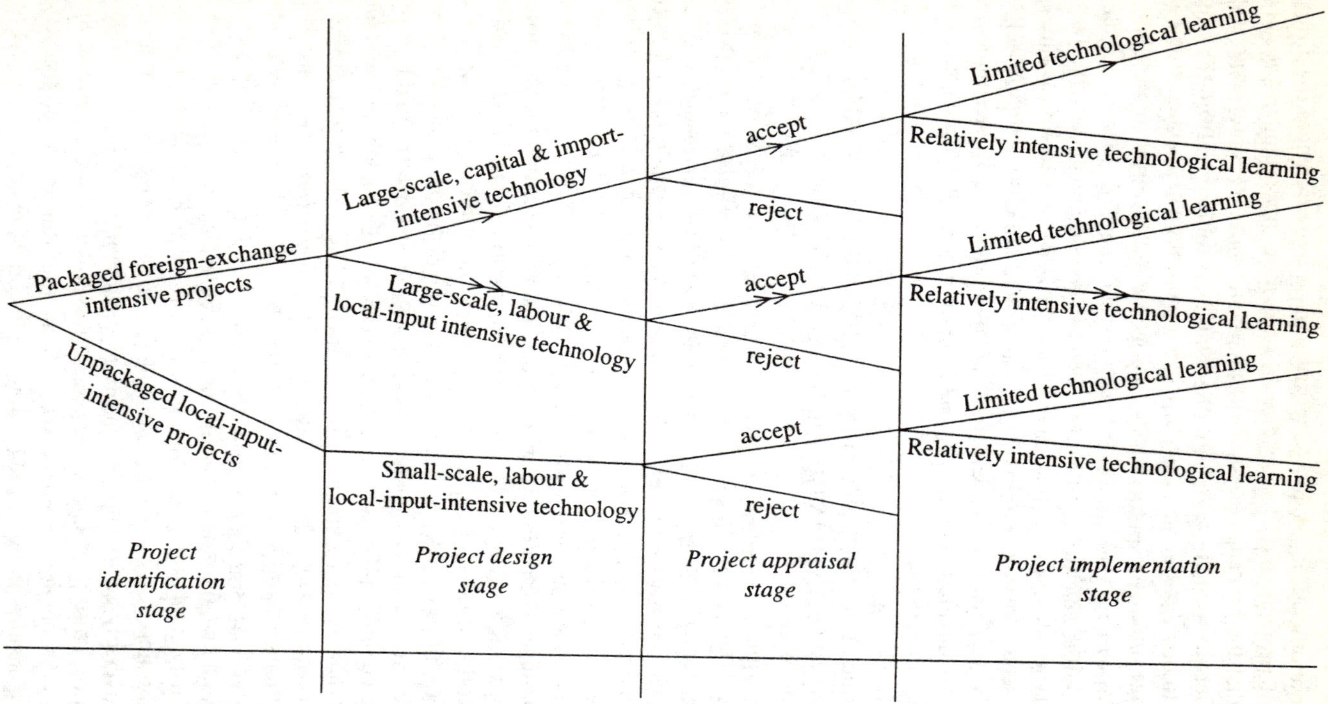

Figure 5.1. A decision-tree representation of the paths to appropriate technology

kind or another. It is usually governed instead by preferences for different types of projects and it is accordingly the most important of the differences that the branches of the first fork should convey. Again on the basis of earlier discussion (especially in Chapters 2, 3 and 4), this requirement seems to be most fully met by branches representing (1) packaged foreign-exchange intensive projects and (2) unpackaged, local-input-intensive projects.

The attraction of the first, we suggested, was such that in the vast majority of cases, the upper branch of this fork would be favoured by public sector institutions. We further suggested that this choice was likely to occasion a particular package of external finance and other inputs: one in which developed market economies are heavily represented, either through aid donors, machinery suppliers, consultants or banking institutions (or through some combination of these). In that these inputs seem to determine closely the technology used in the project, the upper branch of the next fork – representing large scale, capital – and import-intensive methods of production – would again tend to be chosen (albeit at this stage by the foreign collaborators rather than by state institutions in the host country). These very characteristics of the typical project, moreover, may at times tend to increase rather than decrease the probability that the project itself will be approved. (This certainly seemed to happen, for example, with many of the projects in Kenya and Tanzania that were described in Chapters 2 and 3.) And even if this is not the case, it is still rare for projects with these inappropriate technological features to be rejected at the appraisal stage even when there is opposition from one or more of the participating institutions (as was shown in Chapter 3 and is suggested by some other case-studies as well).[41] More often than not, therefore, the project will move on without substantial modification of its technological features to the implementation stage, where, according to the findings presented in Chapter 1, a rather common technological pattern again seems to prevail. In particular, what those findings seem to suggest, is that few incremental technological capabilities generally accrue to public sector institutions during the implementation of highly packaged and capital-intensive investment projects. Such, then, are the various elements that jointly define the stylized technological trajectory shown in Figure 5.1. Obviously, that trajectory will be more likely to prevail the higher the probabilities of the individual events that give rise to it.

Conversely, the second option should also be apparent from the diagram: the lower these probabilities just described at the identification and design stages, the greater the subsequent scope for a more appropriate technology. Consider first the form of appropriate technology that is exemplified by the cases presented earlier in this chapter and which is described in the figure as being large-scale, labour- and local-input intensive. We were at pains to emphasize that these examples did not seem to emerge out of any significant deviation from the goals usually pursued by public sector institutions, nor did they appear to reflect any greater degree of attention than is normally paid by these institutions to the technological aspects of development projects. These cases, we suggested, are better described in terms of how a fortuitous combination of financial or other project inputs happens to favour the lower rather than the upper branch of the second fork in the figure (with the subsequent result that technological capabilities stand a greater chance of being accumulated than is typical of large-scale industrial projects in the public sector).

What has not yet been adequately emphasized, however, is that the 'large-scale labour-intensive project' is not the only way in which appropriate technology might arise in the public sector. One also needs to consider cases where such technology takes the form of numerous small-scale plants. In the case of brick manufacture, for example, 11 labour-intensive plants would be needed to produce the same level of output as one large-scale mechanized operation.[42] In order to produce an amount of bread equivalent to the output of a large-scale automated factory, some 50 small-scale bakeries would be required.[43] Further examples and extensive discussion of this point and its implications are provided in Chapter 8.

These versions of appropriate technology differ from those that have already been described in this chapter (which deviate from the normal pattern at the second fork in the figure), in that they generally need to be considered at the very outset of investment projects if they are to stand any chance of being selected (they need, in other words, to follow from the choice of the *lower* branch of the *first* fork). For, even if only implicitly, biases towards the conception of projects in large-scale terms inevitably seem to preclude later technological options (as indeed is suggested by the way in which the decision-tree itself has been drawn). In effect, what one encounters is in the nature of an irreversibility: once set on a particular technological course, investment projects seldom entertain other possibilities and more rarely still do they actually switch to those possibilities.

In Côte d'Ivoire, for example, the six state-owned sugar complexes constructed during the 1970s 'followed a mass production model centred upon large-scale . . . plantations and vacuum pan (VP) sugar processing techniques. The possibility of experimentation with . . . village plantations or small-scale sugar processing using an open-pan sulphitation technique (OPS) . . . was not considered'.[44] Holtham and Hazlewood make a similar point in relation to a state-owned sugar company in Kenya.[45] And in yet another relevant example, of a large scale state-owned bakery in Tanzania, the option of producing the same amount of bread in 40–50 small-scale units received a 'rather cursory dismissal',[46] instead of serious consideration.

Part of the problem is that the appraisal stage of the project cycle rarely serves the function of generating information about and comparing technological alternatives in public sector investment projects. This observation may be especially true in Sub-Saharan Africa, where the relevant capabilities are often underdeveloped in comparison with many countries in Latin America and Asia, but it also seems to apply to the Third World more generally. Consider for example how the role of project appraisal is perceived by the authors of a detailed study on the technological influence of banking institutions in developing countries. As they see it:

> The appraisal stage is the only phase where one usually finds a discussion of the project's technological features. By that time, however, the project is so far advanced, and so close to a final decision, that no fundamental modification of its technical parameters can be made. Moreover, although appraisal reports usually describe the technical and engineering features of the project, the approach chosen is seldom compared with other possible alternatives, or justified in terms of such comparisons.[47]

Partly on the basis of this observation and partly as a result of their analysis of the project cycle as a whole, Jéquier and Hu conclude – as we do – that the decisive technological influences are usually imparted at the identification and implementation stages rather than when a project is appraised.[48] Yet, as these authors also rightly emphasize, it is the last-mentioned stage that typically receives the most attention in development banks. Not only, for example, do these institutions devote the bulk of their time and effort to project appraisal (accounting in some cases for as much as 80 per cent of staff time), but 'It is also the most glamorous work, and one of the privileged avenues for staff promotion. It has furthermore all the additional attraction that goes with technical complexity and intellectual challenge.'[49]

The comparative neglect of those stages in the project cycle that exert the most powerful technological influence is both a cause and a consequence of the fact that development banks in developing countries seldom seem to conceive of themselves as technological institutions.[50] It is a cause in that activities having a relatively intensive technological component are displaced by activities that have not. It is a consequence in that there is little in the way of a coherent national – or even sectoral – technology policy that might otherwise imbue these institutions with a greater degree of concern with technological issues (including those that are important to the outcome of an investment project). Certainly in Sub-Saharan Africa, as many observers have noted, there has been and for the most part continues to be a rather notable degree of inattention to this type of policy (for reasons that are explored more fully in the next chapter).[51]

These mutually reinforcing institutional constraints make it difficult enough for technological concerns of any kind to be effectively inserted into the project cycle. It is all the more difficult when those concerns relate specifically to appropriate technology, partly because of severe imperfections of information and partly because this technology often takes the form of numerous small-scale plants rather than an alternation in the method of production at a given (large) scale of output. This distinction is important not least because the former version of appropriate technology needs to be considered when a project is first conceived if it is to be introduced at all.

To these already pronounced institutional constraints on appropriate technology, moreover, one has still to add a number of no less intractable problems of a political kind. These will be discussed mainly in Part II.

CONCLUSIONS

This chapter has sought to demonstrate and to explain how appropriate technologies have arisen in the public sector of two African countries, in circumstances that were mostly highly unconducive to these forms of technology. The case-study evidence that was examined for this purpose allows one to draw several important conclusions. The first is that in each case the use of appropriate technology has been accompanied by relatively successful performance at the firm level, thus confirming (and elucidating) the relationship between these variables that was noted in Chapter 1. Significantly, however, these associations did not all take place within one particular institutional setting; on the contrary the case-studies reveal that a variety of different institutions were involved in the promotion of appropriate technology.

What *is* common to most of the examples mentioned in this chapter, though, is a close connection, in one form or another, to developing-country sources of technology or technological knowledge. From a policy point of view, the question that this conclusion immediately raises is how these various connections might be strengthened or extended in the interests of promoting more appropriate industrial technologies in the public sector. In a somewhat different, though related context, Pack, for example, has argued in favour of promoting the employment of developing-country consultants at the feasibility stage of industrial projects, on the plausible grounds that engineers from those countries are 'more familiar with machinery produced in developing countries, or less-expensive equipment available from developed countries'.[52] This proposal also satisfies what, according to the decision-tree, emerges as the key requirement that technological concerns need to be raised during the earliest phases of the project cycle, if they are to have any appreciable influence on the outcome. The same can be said of efforts to find developing-country sources of project finance (in the form, for example, of aid and foreign investment) which, we contend, are more likely than developed-country sources to be associated with the kinds of technologies described above.

Much the same line of reasoning leads one to suspect, moreover, that any such cases of appropriate technology that these policies are able to initiate in one particular country, may also be relevant to other similarly endowed countries in the region. Yet, for this spread effect to be realized in actual practice, technological information will need to be communicated between countries – via an appropriate institutional network of some kind – far more effectively than is typical of current practice.[53] (Indeed, there is already much information of this kind that could usefully be shared between countries in say East or West Africa, but which, because of institutional barriers of various kinds remains confined to individual countries in these regions.[54])

These and other possible mechanisms through which technological linkages to the Third World might be encouraged in the Africa region ought, however, to form part of a more general effort to make development finance corporations and other public sector institutions conceive of themselves as agencies of the state that can and indeed should influence the technological component of investment projects in the industrial sector. In part, and at least initially, this may be a question of providing convincing information about the centrality of technological variables to the development process in general and relatedly, about the benefits that more active technological efforts are likely to engender. For, at least to some extent, the rather striking degree of technological inactivity in the public sector that now exists is less the result of a deliberate rejection of known policies than it is a lack of information about what those policies might entail (and what costs and benefits they are likely to engender). It is unlikely, for example, that policy-makers are aware of the role that technology policy played in the historical experience of the now developed countries, or in the more recent experience of the newly industrialized countries of East

Asia.[55] Partly on this account, policy-making in the public sector is unlikely to take cognizance of either the wide range of technological choice that exists in most sectors of industry, or of the significance that attaches to the acquisition of technological capabilities of various kinds.

Indeed, a similar question can be raised even in relation to far more basic economic principles and policies (including those that are currently being promoted in the name of liberalization in most of Sub-Saharan Africa). Bates, for example, has suggested that:

> Many top policymakers in Africa fail to grasp even the rudiments of economic reasoning: the process by which prices are formed, the role of prices in allocating resources, and the operation of markets. Equally to the point, their constituents are poorly versed in economics. The result is that ill-advised policies get proposed and those who bear their costs fail to oppose them.[56]

Bates goes on to make another, related, observation about the policy-making environment in Africa, that is worth quoting in this regard:

> In Africa there is little economic pamphleteering. The journals and newspapers carry little economic analysis, learned economists are not retained as contributors to the popular press, and economic societies operate sporadically, if at all, and rarely provide forums for assessing the making of government policy.[57]

One of the main conclusions that Bates draws from these observations – that political leaders need to be provided with detailed and well-researched information about the costs and benefits of policy reforms – applies, we suggest, with particular force to the technological realm, where the weaknesses he identifies are likely to be especially pronounced. There is clearly an important role to be played here by international organizations such as the ILO and UNIDO, which have already conducted an impressive amount of research on technology in general and appropriate technology in particular (some of which was undertaken in Africa itself).[58] These organizations should pay particular attention to the mechanisms through which their research findings get translated into actual technology policy and they should serve as active proponents of such policies as those findings may indicate[59] (by ensuring, for example, that research results are widely disseminated in a concise and persuasive form and by means of effective follow-up procedures of various kinds).[60]

If it is to have the maximum possible impact at the actual project level, however, any increased awareness that may result from these and other efforts, needs to be reflected in and accompanied by, rather different procedures from those that currently prevail at this level. For, as this chapter has been at pains to emphasize, those procedures tend to lack a specific technological component and they may in some cases even divert attention from the technological dimension of investment projects.

Much of the problem resides in the role of project appraisal, which, as commonly practised, seldom generates a set of alternative technological options and which usually receives more attention than other stages of the project cycle (such as identification and implementation), that exert a more pronounced (and from the point of view of appropriate technology often an undesirable) influence on technology. If the technological aspects of investment projects are to receive more systematic attention, therefore, at least two rather fundamental changes in existing procedures would seem to be called for. The first demands an increase in the status of and the time invested in pre- and post-project activities and especially those activities that are likely to culminate in the adoption of appropriate technologies and the acquisition of indigenous technological capabilities of one kind or another[61] (the need for identification and support of an indigenous capital goods sector is one of the most prominent examples of what such activities might entail). The second main change requires that project appraisal be made to serve more effectively as a method of generating information about a range of alternative technologies, rather than, as now, of assessing only one or a very limited number of alternatives.[62] (Especially in the case of less well-known and relatively simple technologies, a considerable amount of search and research may be necessary using a variety of different sources, such as 'grey literature', international data-banks and appropriate technology manuals.) As noted in the next chapter, a move in this direction would in fact make social cost-benefit analysis more consistent with its original conception, as a technique designed to choose investments with the highest returns, rather than as a means of approving investments that have, in one way or another, already been pre-selected.

Notes

1. The discussion that follows draws heavily on James (1987).
2. Wells (1983), p. 48.
3. Ranis (1973), p. 402.
4. Ibid.
5. Johnston and Kilby (1975), p. 315.
6. Morehouse (1982).

7. Kaplinsky (1983).
8. Perkins et al. (1977).
9. See for example Frankena (1974).
10. Stewart (1981, p. 97) takes the view that 'Although Third World capital goods industries have been castigated for failing to be "appropriate" and to innovate, on balance they are probably more appropriate than most advanced country technologies; and they may become more so in the future, as innovation is encouraged.'
11. For detailed estimates see for example OECD (1986).
12. See Stewart (1981) and Pack (1982).
13. A World Bank Report on water supply and sanitation technologies, for example, found that though many nonconventional technologies are being used all over the world, there is very little information about them, even among engineers. See Kalbermatten et al. (1980).
14. Okumu (1979).
15. Ibid.
16. Wangwe (1986), p. 16.
17. Barker et al. (1986), p. 123.
18. According to the World Bank (1987) UFI performed best (in terms of DRC measures) among a sample of firms in the engineering subsector.
19. Wangwe (1986).
20. Barker et al. (1986), p. 128.
21. Trait-making as opposed to trait-taking behaviour was described in Chapter 1.
22. Müller (1980) and Wangwe (1992).
23. Ibid.
24. Wangwe (1992).
25. There is a more general point to be made here about the extent of learning from policy experience in the Tanzanian case. According to Hyden (1983) the type of decision-making that has been practised in Tanzania – which he refers to as 'trying to run while others walk' – tends to militate against learning, not just in regard to technology but in many other areas as well. He refers in particular to a disinclination among policy-makers 'to use the past as a source of guidance for the future' (p. 74). Policy mistakes, in Hyden's view, are sometimes recognized but only those of 'large proportions' and change in such cases will only be forthcoming 'if the top leader is ready to accept the mistake' (p. 75).
26. Wangwe (1992), for example, refers to the case of the ZZK farm implements project in Tanzania which began production in 1983. In this case, product design was placed in the hands of Indian experts who introduced unmodified Indian designs; these, however, proved inappropriate to Tanzanian needs and conditions. Wangwe's analysis of this disappointing result leads him to question the competence of the Indian consultants who were responsible for the choice and design of the technology at ZZK.
27. The following description of this case is based on Moorthy (1979). The Government of Kenya has a 33.9 per cent share of Pan African Paper Mills, while another 11 per cent is owned by two Kenyan development finance corporations and the East African Development Bank.
28. Ibid., p. 44.

29. Ibid., p. 47.
30. Barker et al. (1986).
31. Similar results for a sample of privately owned Kenyan firms have been reported in Langdon (1982). He shows how indigenous technological capabilities were acquired by these firms and he suggests that the capabilities thus acquired had a positive influence on firm performance.

    Mytelka (1985) compares Langdon's results with the experience of a sample of state and foreign owned firms in Côte d'Ivoire 'to assess differences in the extent to which the building of indigenous technological capabilities is affected by the ownership or management structure of the firm' (p. 102). Among her main conclusions is that unlike the state and foreign-owned enterprises, the Kenyan firms 'were obliged . . . to be cost-competitive' and this obligation led them to make more efficient (and more appropriate) technical choices (p. 110). Such choices, in turn, 'rendered the imported technology more accessible to local owners and more easily mastered. Over time such firms could grow and deepen their technological know-how' (p. 110).
32. Lall (1987), p. 92.
33. ILO (1992). For a detailed description of the programme see de Veen (1980).
34. Pack (1979).
35. In Kenya the state-owned Mumias Sugar Company is also instructive from this point of view. For, according to Holtham and Hazlewood (1976), Mumias is the most efficient of the sugar complexes in Kenya, partly as a result of technology that, to a greater extent than in the other complexes, was adapted to local conditions. Such adaptations in turn are explained by Holtham and Hazlewood in terms of the wide experience in tropical areas that the managers at Mumias (Booker McConnell Ltd) had already gained.
36. Barker et al. (1986), p. 123.
37. Ibid., p. 83.
38. Ibid., p. 122.
39. Ibid., p. 125.
40. Decision-tree diagrams are commonly used to visualize the outcomes of decisions made in uncertain conditions. See for example the textbook on statistics by McClave and Benson (1985). However, whereas that form of analysis is designed to help identify the optimal decision out of a range of options, our concern is to describe *how* actual decisions occur and to show how decisions and outcomes are connected over time.
41. Gachuki and Coughlin (1988), for example, show that in spite of the enthusiasm expressed in Kenyan policy statements for local-resource and labour-intensive projects, 'some approved projects are the exact opposite of these' (p. 106). They further demonstrate that these approved projects often survived the acute criticism of various government institutions, partly for political reasons, partly because the opposition was fragmented and partly because of administrative weaknesses in the planning process.
42. Kaplinsky (1990).
43. Green (1978).
44. Mytelka (1992), pp. 250–1.
45. Holtham and Hazlewood (1976).
46. Green (1978), p. 18.

47. Jéquier and Hu (1989), p. 82.
48. As we have done, Jéquier and Hu pay a great deal of attention to the project identification stage and to the bias in favour of large-scale projects that often occurs at this stage. They also emphasize that this bias has important technological implications since 'identifying a project at the outset as a large project will in many cases govern the subsequent choice of technology, or at least rule out certain technological options'. See Jéquier and Hu (1989) p. 60. Where our analysis differs from theirs, however, is with respect to the causes of the scale bias at the identification stage. For them:

> perhaps the most important reason for this tendency is to be found in the planning process itself and in the [development] bank's economic studies. Both types of work . . . are by essence macro-economic, and focus on the medium and long term. These two features almost inevitably lead to 'thinking big' when it comes to project ideas . . . By looking seven to ten years ahead (this is the horizon for the next five-year plan, or the bank's country studies), the planner and the banker will almost inevitably come to think of a new production capacity which will match the consumption gap anticipated at that particular point in the future. *And this new production capacity, precisely because it is situated relatively far into the future, will correspond to a fairly large plant* (p. 61; emphasis added).

Our analysis in contrast focuses on the *failings* of the planning system; on the predominance of *short* rather than long-run concerns; on the related political and other motivations of the bureaucratic class and on conflicts within the state itself. (These last issues are discussed in Chapters 7 and 8).
49. Jéquier and Hu (1989), p. 176.
50. For a detailed discussion of the way that development banks in developing countries typically define their functions, see ibid.
51. In some of the larger, more developed countries of Asia and Latin America, in contrast, development banks are more given to the consideration of technological issues. It is true that these concerns have not often reflected themselves in the form of appropriate technology; nevertheless, the notable case of the Swaraj tractor in India illustrates what can be achieved by a development bank that defines itself in these terms and which is able to pose the relevant technological questions at an early stage of the project cycle. The Swaraj case is all the more remarkable when one considers that the Industrial Development Bank of India:

> in the face of a certain indifference on the part of several government agencies for this indigenous innovation and despite the preference of other agencies for the importation of foreign tractors or the manufacturing of tractors under foreign licence, deliberately sought to promote the development and production of the indigenously-designed Swaraj tractor despite the fact that it was only at the prototype stage. This very supportive attitude towards indigenous technology can be traced back to the bank's explicit policy to play not only the role of a development bank, but also that of an institution responsible for promoting indigenous innovation (Jéquier and Hu, 1989, p. 70).

52. Pack (1986), p. 172.
53. There are numerous proposals for international information networks in the area of technology in Bhalla (1979).
54. The proposed Inter-Agency African Regional Network for farm tools and equipment, for instance, was based partly on this recognition (See ILO, 1986).
55. As expounded in relation to the latter group of countries by, for example, Pack and Westphal (1986).
56. Bates (1988), p. 129.
57. Ibid., p. 130.
58. See for example the list of publications contained in ILO (1992).
59. The Technology and Employment Branch of the ILO itself commissioned a report to examine how its research programme could be made to exert a greater degree of influence on the policy-making process in developing countries. See ILO (1992).
60. Ibid. For a useful, more general, discussion of the mechanisms through which economic research is translated into public policy outcomes, see Nelson (1987).
61. For a detailed discussion of this proposal in the specific context of development finance corporations in developing countries, see Jéquier and Hu (1989).
62. Ibid., and Stewart (1979).

References

C. Barker, M. Bhagavan, P. Mitschke-Collande and D. Wield (1986). *African Industrialization*, Gower.

R. Bates (1988). 'The Political Basis for Agricultural Policy Reform', in S. Commins (ed.), *Africa's Development Challenges and the World Bank*, Lynne Rienner.

A. Bhalla (ed.) (1979). *Towards Global Action on Appropriate Technology*, Pergamon.

M. Frankena (1974). 'The Industrial and Trade Control Regime and Product Designs in India', *Economic Development and Cultural Change*, vol. 22(2).

D. Gachuki and P. Coughlin (1988). 'Structure and Safeguards for Negotiations with Foreign Investors: Lessons from Kenya', in P. Coughlin and G. Ikiara (eds), *Industrialization in Kenya*, Heinemann.

R. Green (1978). 'The Automated Bakery: A Study of Decision Taking Goals, Processes and Problems in Tanzania', Institute of Development Studies, Sussex University, Working Paper no. 141.

G. Holtham and A. Hazlewood (1976). *Aid and Inequality in Kenya*, Croom Helm.

G. Hyden (1983). *No Shortcuts to Progress*, Heinemann.

International Labour Office (1986). *Excess Manufacturing Capacity and Farm Equipment Needs in Kenya*, Geneva.

International Labour Office (1992). *The Impact of World Employment Programme Research on Technology*, Geneva.

J. James (1987). 'Population and Technical Change in the Manufacturing Sector of Developing Countries', in D. Gale Johnson and R. Lee (eds), *Population Growth and Economic Development: Issues and Evidence*, University of Wisconsin Press.

N. Jéquier and Y. Hu (1989). *Banking and the Promotion of Technological Development*, Macmillan.

B. Johnston and P. Kilby (1975). *Agriculture and Structural Transformation*, Oxford University Press.

J. Kalbermatten et al. (1980). *Appropriate Technology for Water Supply and Sanitation: A Summary of Technical and Economic Options*, World Bank.

R. Kaplinsky (1983). *Sugar Processing: The Development of a Third World Technology*, IT Publications.

R. Kaplinsky (1990). *The Economies of Small*, IT Publications.

S. Lall (1987). 'Long-Term Perspectives on Sub-Saharan Africa', background paper, Special Office for African Affairs, World Bank.

S. Langdon (1982). 'Indigenous Technological Capacity in Africa: the Case of Textiles and Wood Products in Kenya', Paper prepared for International Workshop on Facilitating Indigenous Technological Capacity, Edinburgh University, May.

J. McClave and P. Benson (eds) (1985). *Statistics for Business and Economics*, 3rd edn, Dellen Publishing.

V. Moorthy (1979). 'The Experience of a Developing Country of Africa', in UNIDO, *Appropriate Industrial Technology for Pulp and Paper Products and Small Pulp Mills*, Monographs on Appropriate Industrial Technology, no. 3.

W. Morehouse (1982). 'Opening Pandora's Box: Technology and Social Performance in the Indian Tractor Industry', in F. Stewart and J. James (eds), *The Economics of New Technology in Developing Countries*, Frances Pinter.

J. Müller (1980). *Liquidation or Consolidation of Indigenous Technology*, Aalborg University Press.

L. Mytelka (1985). 'Stimulating Effective Technology Transfer: the Case of Textiles in Africa', in N. Rosenberg and C. Frischtak (eds), *International Technology Transfer: Concepts, Measures and Comparisons*, Praeger.

L. Mytelka (1992). 'Ivorian Industry at the Cross-Roads', in F. Stewart, S. Lall and S. Wangwe (eds), *Alternative Development Strategies in SubSaharan Africa*, Macmillan.

R. Nelson (1987). 'The Economics Profession and the Making of Public Policy', *Journal of Economic Literature*, vol. XXV.

OECD (1986). *Development Co-operation*, Paris.

J. Okumu (1979). 'Foreign Relations: Dilemmas of Independence and Development', in J. Barkan and J. Okumu (eds), *Politics and Public Policy in Kenya and Tanzania*, Praeger.

H. Pack (1979). 'Technology and Employment Constraints on Optimal Performance', in S. Rosenblatt (ed.), *Technology and Economic Development: A Realistic Appraisal*, Westview Press.

H. Pack (1982). 'The Aggregate Implications of Factor Substitution in Industrial Processes', *Journal of Development Economics*, vol. 11.

H. Pack (1986). 'The Technological Impact of World Bank Operations', in R. Feinberg (ed.), *Between Two Worlds, The World Bank's Next Decade*, Overseas Development Council.

H. Pack and L. Westphal (1986). 'Industrial Strategy and Technical Change: Theory vs. Reality', *Journal of Development Economics*, vol. 22(1).

D. Perkins et al. (1977). *Rural Small-Scale Industry in the People's Republic of China*, University of California Press.

G. Ranis (1973). 'Industrial Sector Labor Absorption', *Economic Development and Cultural Change*, vol. 21(3).

F. Stewart (1979). 'A Note on Project Appraisal and Appropriate Technology – Some Suggestions for Further Work', mimeo, Queen Elizabeth House, Oxford.

F. Stewart (1981). 'International Technology Transfer: Issues and Policy Options', in P. Streeten and R. Jolly (eds), *Recent Issues in World Development*, Pergamon.

J. de Veen (1980). *The Rural Access Roads Programme – Appropriate Technology in Kenya*, ILO, Geneva.

S. Wangwe (1986). 'Technology Imports, Technological Learning and Self-Reliance in Tanzania', World Employment Programme Research, Working Paper no. 165, ILO, Geneva.

S. Wangwe (1992). 'Building Indigenous Technological Capacity: A Study of Selected Industries in Tanzania', in F. Stewart, S. Lall and S. Wangwe (eds), *Alternative Development Strategies in SubSaharan Africa*, Macmillan.

L. Wells (1983). *Third World Multinationals*, MIT Press.

World Bank (1987). *Tanzania: An Agenda for Industrial Recovery*, vol. II: Annexes and Statistical Appendix, Washington, DC.

6 Foreign Aid and the Bureaucratic Choice of Techniques

So far our analysis has focused primarily on public sector institutions in the industrial sector, though we have at times referred also to the important role played by a number of other, mostly developed-country, institutions. Among these other institutions, foreign aid donors would appear to warrant particular attention in view of the striking extent to which they have financed public investment in most African countries south of the Sahara (as Chapter 1 explicitly sought to show). And as its title suggests, the purpose of the present chapter is to incorporate foreign aid donors more systematically into our analysis. So doing, we suggest, takes two distinct though related forms.

The first is by extension: by means, that is to say, of an argument that aid-donor motivation – and hence much of its behaviour – can be understood in terms similar to the bureaucratic processes already discussed in relation to public sector institutions in several African countries. The second way in which aid-donor institutions are incorporated into the foregoing analysis is based on the recognition that the technological influence they appear to exert so systematically cannot be divorced from the behaviour that parastatals and other state institutions *themselves* exhibit. Indeed, it often seems to be the case that this influence is reinforced rather than ameliorated by the various institutions of the African state, which, directly or indirectly, bear on the technological behaviour of the public sector.

AID DONOR MOTIVATION IN AFRICA: THE 'MOVING MONEY' SYNDROME

Much of our foregoing analysis of public sector institutions can be described as Hirschmanian in that it searches for the 'hidden rationalities' underlying the paradoxical technological behaviour of these institutions.[1] These 'rationalities', we suggested, were 'hidden' from the purview of most traditional approaches to the choice of technology in developing countries. Moreover, in so far as their consequences are not apparent without fairly close technical scrutiny, the same 'rationalities' tend to be hidden from a policy point of view as well (recall in this connection the observation from Chapter 2, that inappropriate technical choices manifested themselves to leading policy-makers only when acute mechanical deficiencies occurred).

Although it is not described in these terms, Tendler's analysis of foreign aid institutions also proceeds along distinctly Hirschmanian lines.[2] For its main goal is to uncover the 'hidden rationalities' that underlie paradoxical behaviour of the following kind:

> It appears to anyone who has spent much time in a development assistance organization that people are making decisions as if aid funds were abundant. How can this be, when everyone knows how truly scarce development assistance is? How can this happen, when the borrower country is constantly facing problems caused by foreign exchange and capital scarcity?[3]

This paradox can be resolved, according to Tendler, by means of the recognition that aid scarcity and abundance both exist, but in different contexts. From the purview of the macro-economist whose task is to estimate the foreign capital needed to meet policy goals of various kinds, it is all too obvious that there is a scarcity of foreign aid (as expressed, perhaps most commonly, in the form of 'gaps' of one sort or another). From the micro-economic perspective of the donor organization, in contrast, the supply of aid funds:

> is perceived in relation not to total estimates of need and supply but to the amount of work or time required to commit the funds available. By job definition, the member of a development assistance organization is charged with finding worthy projects for the funds available. His perception of the supply of these funds is related to the availability of financeable projects. Because long delays and considerable work accompany the putting together of project proposals, there may be fewer financeable projects, at any given moment in time, than funds available, From the employee's point of view, the scarce commodity is frequently the project rather than the funds.[4]

Though Tendler does not relate this discussion of aid 'abundance' to particular parts of the Third World, it would seem especially applicable to the circumstances prevailing in most of Sub-Saharan Africa. On the one hand, aid-flows to this region have been remarkably high not only in relative but also in absolute terms (In the mid-1980s, for example, the ratio of total ODA receipts as a percentage of GNP was 5 per cent for Sub-Saharan Africa as against a figure of around 1 per cent for developing countries as a whole; in absolute terms Sub-Saharan Africa received more than 30 per cent of total ODA, although its share of the total population in the Third World is only some 11 per cent).[5] On the other hand, whether it is defined according to technological capabilities, labour skills, infrastructure or some other related measure, 'absorptive capacity' in Sub-Saharan Africa is typically below the Third World average and not infrequently considerably so.[6] To a greater extent than elsewhere, therefore, viable (or 'aidworthy') projects in this region are likely to be hard to come by.

The combined effect of these two factors is to make improbable a view of aid donor institutions operating in Africa as having to ration scarce funds among a large number of competing project claimants. What becomes far more credible is the opposing view expressed by Tendler that:

> the supplier of funds, the assistance institution, has to take over part of the activities typically carried out by the interested consumer: before the supplier can 'sell' his funds, he must first help to generate demand for them. . . . much of the [development] agency's staff time is spent drumming up future business-exploring possibilities for projects with borrower governments, suggesting ways of preparing project applications, or making its own evaluation of sectors in which conditions for lending are favorable . . . and then looking closely at those sectors for possible projects. . . . Project identification and preparation amounts to a generation of demand for aid funds.[7]

The more resources are seen in this way – 'as an inert mass which will not move without great effort'[8] – the more achievement within the aid donor institution is likely to be appraised according to how much money is 'moved'. The more, that is to say, the mere *quantity* of resources transferred, without regard to considerations of effectiveness, becomes the yardstick of accomplishment within the donor institution. And, given the common practice of financing only import, or foreign exchange costs, the more will the bias in favour of large-scale, capital and import-intensive technologies tend to be, since it is projects with these features that most readily permit money to 'move'.

In Africa, as elsewhere in the Third World, this form of bias may have been most pronounced during the 1970s when aid funds seemed to be especially plentiful, although the 'moving money' phenomenon continues to be observed in both bilateral and multilateral aid institutions. A recent paper by Gauthier, for example suggests that 'lending objectives' are the most common performance targets in lending institutions such as the World Bank. In particular, in 'the organization, a department or specific officers are required to meet certain disbursement targets for specific regions, country or sectoral in recipient countries. Promotions and bonuses are grounded on the achievement of these targets'.[9] Even at its most pronounced, however, the technological bias that is imparted by the need to commit and disburse aid funds operates subtly and somewhat invisibly in the African context, just as it does in the case of bureaucratic decision-making in the public sector. For in both cases, what usually occurs is not that alternative projects and technologies are considered and rejected. The problem is rather that *these alternatives remain largely out of sight, bypassed by a decision-making process that, in donor and recipient institutions alike, provides little or no incentive to deviate from the biased outcomes we have described.* And whereas, in some developing countries with relatively well-developed capital goods sectors, these rejected alternatives may show a 'voice',[10] in Sub-Saharan Africa this is usually much less true. For in most of the countries in this region 'the neglected domestic supplies, the simpler technologies, or the smaller projects would be quite difficult to find. They would not be embodied in bypassed persons, groups or industries, and they would not have the ardent spokesmen that a developed manufacturing sector has'.[11] This is an extremely important observation, to which we shall return later, when dealing with the political economy aspects of technological and other behaviour in the public sector.

Many of the general observations made in the previous paragraph can be observed at the level of specific projects involving aid donors on the one hand and public sector institutions on the other. Consider, for example, the circumstances in which an automated bakery was established by the Tanzanian government with the assistance of Canadian aid. Close scrutiny of this case reveals that an alternative to the automated methods might conceivably have taken the form of some 40–50 small-scale bakeries, using labour-intensive machinery produced in the informal sector.[12] Yet, in spite of its simplicity and its comparatively low capital and import costs, this alternative apparently received only very brief consideration. Green suggests that while this 'rather cursory dismissal' was defensible on a number of more or less reasonable grounds,

> it remains a clear weakness in the . . . process that no serious effort was made to see whether there was any interest in hand baking by informal sector artisans, no attempt to set up one or two test units and no attempt to interest NMC [National Milling Corporation] in providing

procurement/marketing services and advice to the initial cooperative hand bakeries to see what could be achieved.[13]

Dominated as it is by large-scale, state-owned enterprises, the sugar industry in Kenya is also instructive from these various points of view. As in the Tanzanian case that has just been described, relatively small-scale, labour-intensive alternatives have been neglected in Kenya in spite of the considerable capital and foreign-exchange savings that they afford in comparison with the large-scale, vacuum-pan technologies adopted by the state-owned enterprises.[14] And again as in the Tanzanian case, it seems more than likely that this pattern of production has been heavily influenced by the availability of foreign aid. The Mumias Sugar Company, for example, has been described as an attractive project from the British donor's point of view, among other reasons because it was 'large and highly visible . . . and as it included a factory it had a high import content'.[15] This cheap foreign exchange, in turn, may have 'made it harder for the Kenya government to assess objectively costs and benefits of the different methods'.[16] Certainly, few other techniques were considered for Mumias than the one that was proposed by its British managing agents, Booker McConnell. Similarly, for the case of hand-pumps in Kenya, it has been argued that foreign aid perpetuates a dependence on imported products because it neglects the existence of local capabilities and machinery for producing these same products.[17]

This neglect of the more appropriate technology and of local capabilities may also have reflected a lack of political influence on the part of those in Kenya who would have gained therefrom. Kaplinsky, for example, appears to have this point in mind when he concludes his analysis of appropriate technology in a number of different sectors in several African countries (including sugar in Kenya) with the observation that,

> unlike Asia, Africa seems to lack a domestic political constituency whose interests are extended through the choice of small-scale technology. For this reason the panoply of state support and price interventions tends to be geared towards the mobilization of large-scale inappropriate investments, despite their demonstrated failure. In these environments the idea of AT [appropriate technology] is often received with some hostility. By contrast, in many parts of Asia a ready constituency emerges in support of small-scale industry.[18]

We shall return to this theme in a later chapter that deals directly with the political economy of centralized versus decentralized modes of organizing different forms of industrial technology in Africa.

THE GENERAL LACK OF COUNTERVAILING INFLUENCE

What we have suggested so far in this chapter is that the behaviour of aid donor institutions in Sub-Saharan Africa is and has been especially heavily influenced by the need to 'move money' (a need, which, according to some observers pervades all such institutions to a greater or lesser extent). We have not yet examined, however, the various forces that might exert a countervailing influence over this behaviour. These influences emanate both from within the donor institutions themselves and from public sector institutions in the recipient country.

Project Appraisal by Donor Institutions

One might reasonably suppose that the aforementioned biases to which foreign aid seems to give rise, would be limited at the appraisal stage of the project cycle; in particular, one might expect social cost–benefit analysis to exert a countervailing influence against the bias towards large-scale, capital and import-intensive technologies, that we suggested might be especially pronounced in Sub-Saharan Africa.

For what this form of analysis explicitly seeks to achieve is the substitution of (usually 'distorted') market prices with 'shadow' prices that reflect social objectives and the true opportunity costs of resources. Thus,

> the market wage rate may (e.g. because of Trade Union pressure) be much above the actual opportunity cost of labour (or what the worker would be producing in alternative occupations). Moreover, the government may place considerable priority on increasing employment. Hence SCB [Social Cost Benefit Analysis] would use a shadow wage below the market wage and thus secure more labour-intensive projects than if the market wage were used.[19]

Similar adjustments to the market rate of interest (which may often be too low in relation to its shadow value) and the exchange rate (which is usually overvalued), lead one to expect that, with other things being equal, SCB would favour labour and local-resource-intensive projects and discourage projects that depend heavily on capital and other imported inputs.

In practice, however, this countervailing influence tends to be undermined by a variety of factors, some of which influence the *extent* to which SCB is used, while others influence the *manner* of its use by aid donor institutions. The former limitation bears, for example, on both bilateral donors, whose use of SCB is by no means universal, and on multilateral donors, such as the World Bank, only the direct part of whose total investments in developing countries undergo rigorous SCB analysis. Other investments, as noted earlier, are undertaken indirectly by development finance corporations with the loans that the Bank makes to them. These investments, it seems, do not always, or even generally, undergo project appraisals based on standard SCB methods,[20] while the Bank's influence over the technological aspects of DFC projects also appears to be highly limited. Indeed, with regard to this last point, the Bank itself acknowledges that while the

> Technical aspects of subprojects above the free limit are reviewed . . . the possibilities for significant changes are severely constrained. The Bank reviews the justification for the technology selected, but leaves the actual decision to the DFC and the entrepreneur. . . . Since Bank funds mostly cover the foreign exchange components of subproject costs, the Bank has little influence in reducing reliance on imported technologies that are capital intensive.
>
> The Bank's review of the proposed technology is also limited in its detail because feasibility studies are normally not included when a subproject is sent to the Bank for approval. The Bank in any case would not be familiar with the specific conditions which exist in the field.[21]

Even where standard SCB methods are used, however, the degree of countervailing influence that is thereby exercised on behalf of appropriate technology is frequently very limited and may at times even be counterproductive (in that, as some observers have argued, the resources devoted to project appraisal and rate of return analysis divert attention from other crucial dimensions of project performance including those of a technological kind).[22] That such influence is so limited reflects mainly the typical use of SCB as a means of evaluating a *given* project rather than as a means of identifying and preparing a *range* of competing projects. Of the numerous sources that could be cited in support of this proposition,[23] an article about the World Bank by Nathaniel Leff is perhaps the most telling since it is in that institution that one would most expect to find SCB analysis used according to its original intention of optimally allocating a given amount of investment resources. Yet, according to Leff, this is not at all how SCB analysis is typically conducted in the Bank; it is employed there rather as a means of ensuring that selected projects yield the required minimum rate of return.

> That use, however, is very far from the purpose or the spirit of SBCA [social benefit–cost analysis]. The methodology was designed for

> choosing the investments with the highest returns, not those that were pre-selected on other criteria but pass minimum standards. One objective of the Appraisal Reports and their use of project rates of return is apparently to lend legitimacy to the Board's decisions. . . . Selection of investments from a range of alternative project proposals, however, is what investment choice is all about. Consequently, for SBCA to be denied a major role in that process is a distinct downgrading for the technique.[24]

If project identification were not itself subject to biases in favour of one type of technology over another, this 'downgraded' version of SCB analysis would not discriminate systematically against appropriate technology. Since, however, the earliest phases of investment projects tend to favour large-scale, capital-intensive projects (as we have repeatedly argued), the form of SCB analysis described in the previous paragraph *does* effectively discriminate in this way because it exerts no countervailing influence in the opposite direction (as would be the case for example with a form of SCB analysis that sought explicitly to overcome what are often described as the pronounced information biases against appropriate technology).

Even in its diminished role as a guarantor of minimum project standards, moreover, SCB analysis has not been able to prevent some of the most conspicuous cases of inappropriate industrial technology in Sub-Saharan Africa (This much at least one might have expected, given what we know from Chapter 1 is the negative influence of large-scale, capital and import-intensive technology on economic efficiency measured in both market and shadow prices). In the disastrous Morogoro shoe project in Tanzania, for example, the high degree of capital intensity was justified in the World Bank's project appraisal report by the quality demands of the export markets in which the product was to compete (as discussed in Chapter 3). In the event, however, the planned exports grossly failed to materialize, with the result that the *ex post* rate of return on the project is far below what was anticipated at the time of appraisal.

Another Bank project in Tanzania that is often cited in this context is the pulp and paper mill at Mufindi, which the Bank's own appraisal report described as 'expensive in relation to the economy of the country'.[25] So expensive in fact was the mill, that its investment cost per ton of installed capacity in 1978 (of 4200 US dollars) exceeded by a factor of 2.3 the highest figure then prevailing in the industrialized countries.[26] In this case, the size and capital intensity of the project were justified largely on the grounds of economies of scale, which, again in the Bank's own words precluded 'simpler and less versatile concepts' from consideration.[27] As in the shoe project, however, this meant that the outcome of the endeavour was made to depend rather heavily on the availability of markets, which, again as in that case, mostly failed to materialize.[28]

To a certain degree, the overestimates of demand in these two cases might have reflected factors that were difficult or impossible to foresee at the time of appraisal. On the other hand, however, the extent of the overestimation (particularly in the Morogoro case) was such as to make one suspect that something of an 'optimistic bias' may have been at work on the benefits side of the SCB calculus.[29] Indeed, one observer of the Bank's internal review procedures, Howard Pack, has gone so far as to suggest that this bias is 'inherent',[30] the direct result, he believes, of pressure to achieve high target rates of lending. In different guise, then, the very same need to commit aid resources that is encountered at the project identification stage as a bias in favour of particular technologies, may reappear during appraisal to render less effective the opposition to those same technologies.

The Lack of Countervailing Influence in Recipient Country Institutions

To the extent that the 'bureaucratic-man' mechanism operates in public enterprises in Sub-Saharan Africa, the goals and procedures of aid donor institutions that have just been described, would encounter little opposition. Indeed, one can probably then speak more accurately of a close *coincidence* of interests between the parties involved in public sector investment projects (in that the desire of the donor institution to spend foreign exchange is matched on the recipient side by an equally powerful drive to acquire foreign exchange). Public enterprises are not however the only institutions in the public sector that interact with aid donors. It is important therefore to examine why these other institutions have apparently exerted so little influence over aid donor behaviour in general and their technological behaviour in particular. Part of the answer to this question draws on evidence presented (albeit sometimes only rather briefly) in previous chapters; other parts however depend on arguments that need to be newly advanced.

THE LACK OF AN OBJECTIVE FUNCTION

It is commonly observed that the institutional framework within which public investment in Sub-Saharan Africa occurs lacks a clear statement of government priorities (as opposed to a mere listing of which goals are regarded as desirable). What is less commonly observed however is that it is precisely this lack of a national objective function which invests aid with so much influence in the recipient countries. One of the few exceptions to this general observation is Holtham and Hazlewood's study of foreign aid in Kenya, which also makes clear that the problem of specifying an objective function is in essence a political one. They observe in particular that:

> What donors frequently diagnose as a technical failure ('inability to draw up adequate sectoral plans') for which they prescribe technical assistance is in fact a political failure. Priorities in development or clashes between ministries over policies or for resources are all too rarely resolved in Kenya by Cabinet decision. Cabinet meetings for that kind of business are rare. . . . Hence some political decisions are taken at administrative level and more are not taken at all. . . . This lack of political direction is why the Kenya Treasury does not rank development projects. That, too, is wrongly viewed by some donors as a technical failure. . . . [The Treasury] does not have Cabinet decisions saying this ministry's baby before that one. Without that, each ministry is a semi-independent fief of its top officials called to render particular service now and then when the President speaks.[31]

The political vacuum thus created gives rise to a tendency for foreign aid to become a decisive element in project design and selection. For what occurred in practice in Kenya during the 1970s seemed to be that:

> operational ministries after initial talks with a donor, at which the Treasury may or may not be represented, often pass proposals to the External Aid Division for forwarding 'officially' to the donor without the Planning Division being told at all. . . .
>
> Once the donor has agreed funds it becomes very hard for the Treasury to refuse a budgetary allocation for the project, which is a *fait accompli*. There is a rule that Treasury officers should always be present for negotiations on revenue and finance between an operational ministry and the donor. Even if this obligation is carried out, there is no ruling that Treasury representatives have to be present for technical talks. More-or-less technical talks may be the important early ones, however. In any case, the technical talks always have enormous implications for finance. . . . In order to be able to meet obligations which it may be confronted with, the Treasury is reluctant to make a budgetary allocation for any

> project unless there is a prospective donor. *Donors therefore . . . undermine the functions of the Treasury and Planning Division. Operational ministries do not have to go to the Treasury and compete for funds on any rational basis. If they try that, however technically good a project may be, they are unlikely to get a cent. The first step therefore is usually to find a donor.*[32]

I have quoted from this text so extensively because it provides an unusual degree of insight into the institutional processes that invest donor preferences with such importance. It also helps to explain why SCB analysis is not widely used in Kenya's public sector institutions, or, more generally, why in spite of enormous efforts to develop project appraisal techniques 'relatively few countries [in Sub-Saharan Africa] have systematically established procedures or institutional capacity for economic project appraisal at either sectoral or core ministry level'.[33]

Not all African countres, however, have experienced the types of institutional problems with aid donors described so clearly in relation to Kenya by Holtham and Hazlewood. Botswana, for example, appears to have been able to use aid as a contribution to well-defined priorities rather than, as in the more typical Kenya case, having the influence of aid strengthened by the lack of an objective function at the project, sectoral or national level. Part of the success that the former country enjoyed from this point of view, has been ascribed by some commentators to the way in which its planning priorities were formulated. In particular, Botswana's development plans 'were not prepared in a separate planning commission, detached from the rest of government, but jointly by the planners and the executing departments. Consequently, the resulting sectoral programmes had an operational reality that donor agencies appreciated'.[34] (More generally, the comparatively successful performance of the public sector in Botswana seems to reflect, among other factors, a deliberate policy decision to keep parastatals down to a manageable number; in so doing policymakers resisted the tendency to over-expansion, which has beset almost all other countries in the region.[35])

As has recently been described in a World Bank publication on structural adjustment in Africa, Ghana's experience in the mid-1980s represents another and somewhat similar example of how the management of foreign aid can be improved by institutional reform in the recipient country. First of all, by scaling their public investment programme down to a relatively small number of core projects – 100 in the first instance and 29 thereafter – the Ghanaians were able to make that programme far more consistent with the organizational and other capabilities that were then available.[36]

Furthermore, the process of identifying the elements of a core programme may itself have improved the management of foreign aid in that country. The reason is that 'Core programs push governments to set priorities for spending and to design more effective mechanisms for channelling domestic counterpart resources to important foreign aid-financed projects.'[37] It is not difficult to imagine that in these various ways the core investment programme may have made it easier for the country to 'meet its most critical investment objectives' as the World Bank report suggests.[38] What is not at all clear from that report, however, is the particular combination of institutional and political factors that set the Ghana case apart from the numerous other African countries, where the outcome of core programming exercises has been very much less successful.

DONOR PROLIFERATION AND INSTITUTIONAL DESTRUCTION

The main purpose of the previous section was to show how institutional and political failings in one African country invested aid donor preferences with more significance than might otherwise have been the case. It is also important to recognize, however, that the causality may run in the other direction as well; in particular, foreign aid *itself* may contribute to weaknesses in the very institutions that might otherwise have been able to exert a countervailing influence on behalf of the host country.

Morss, for example, has suggested that the proliferation of aid donors and projects has given rise to 'institutional destruction' in Sub-Saharan Africa.[39] By this he seems mainly to mean that government institutions have become more concerned with meeting the disparate and demanding requirements of a wide range of donors (numbering as many as 69 in some countries), than with discovering and testing their own conceptions of what should be done. What is 'destroyed' by donor and project proliferation, therefore, is essentially the process of experimentation on which institutional – no less than individual – learning so crucially relies.[40] Institutional destruction may also occur, however, in a more specific technological sense than this argument suggests. For one thing, the technological diversity that usually accompanies donor proliferation tends to inhibit local spare parts production and to this extent robs local industry of an important opportunity to acquire the relevant technological capabilities. In addition, the more marked is this technological diversity, the more limited tends to be the applicability of the capabilities that are acquired in relation to the use of any one technology.[41] To this extent, therefore, donor proliferation may also be described as inhibiting the development of user capabilities in public sector institutions.

It is only fair to acknowledge, however, that donor proliferation is not the only source of these unfortunate effects – government policy failures in some African countries also have to bear part of the responsibility. In Kenya, for example, the government could have used the Central Tender Board to reduce the excessive numbers of different makes and models of products such as automobiles and hand-pumps.[42] Yet, according to Ikiara, the Board has not been used for this purpose, partly, it seems, because of a lack of coordination between the relevant state institutions.[43]

THE LIMITED INFLUENCE OF SCIENCE AND TECHNOLOGY INSTITUTIONS

Previous chapters have suggested that neither development finance corporations nor parastatal holding companies have viewed their activities from a specifically technological perspective. We have also seen that these institutions were frequently able to use external project finance as a means of eluding some of the most important regulatory agencies of the state. These institutional 'facts of life' meant that even a well-functioning system of science and technology institutions would have found it difficult to influence decision-making processes in public sector investment projects. Beset as they mostly have been by a variety of acute organizational and political difficulties, the growing numbers of scientific and technological institutions in Africa have confronted a task that was thereby made all the more difficult.[44]

National institutions for technology policy were (and for the most part still are) at a particular disadvantage from this point of view, in that they have often tended to be given

> responsibility without authority: their statutes give them a broad mandate without conferring upon them the power or resources they need to act. . . . Were they to develop a national technology policy – and . . . few ever get that far – they would be unable to carry out its specific action programs.[45]

The African experience is of course far from unique in this respect; indeed, as Girvan and others have pointed out, it is difficult to identify any cases where national institutions have had a discernible impact on actual technological decisions.[46] Unfortunately, however, because these same institutions tend to give the distinct impression that something is actually 'being done' about science and technology activities in developing countries',[47] the problem only tends to perpetuate itself.

Though they have been somewhat less intensively studied than their national counterparts, technology institutions in the industrial sector in most African countries have also been afflicted by organizational and other problems that, one suspects, bore rather heavily on their ability to influence the technological outcomes of public sector projects (I say 'suspects' because there are hardly any detailed case-studies that can confirm this particular proposition). In Tanzania, for example, there are numerous institutions in the industrial sector that are directly concerned with technological issues of various kinds, ranging from research and design to consultancy services. Yet, as Wangwe and others have shown, the activities of these institutions have been too hampered 'with practical problems of coordination and consolidated linkage effectively to link technology import policy to technological learning and self-reliance'.[48] In Kenya, an equally wide range of technological functions in the public sector has been delegated to the (industrial) sector level. Again, however, while some degree of success has been achieved at this level – most notably by the Kenya Industrial Research and Development Institute – institutional constraints of one kind or another remain all too obvious.[49]

CONCLUSIONS

The first part of this chapter has sought to argue that the motivation of foreign aid donors can be described in terms that are similar to the bureaucratic processes discussed in Chapters 2 and 3 in relation to public sector institutions. To this extent, we suggested that one can properly speak of a coincidence rather than a conflict of interests between the domestic and foreign parties involved in public sector investment projects.[50] The nature of these convergent interests, we further suggested, is often such as to impart a distinct, if implicit, technological bias to investment projects in the industrial sector. The second part of the chapter attempted to show that although there are numerous ways in which countervailing influences could be exerted against this bias, in donor as well as recipient institutions, in practice these forces are usually not very significant (though we did take note of the fact that at least a few countries appear to have been relatively successful in improving the management of aid by institutional reforms of one kind or another).

From a policy point of view, what most obviously needs to be changed is the organizational environment within aid donor institutions, which creates the impression among its staff-members of an 'abundance' of aid. 'By significantly reducing the number of decisions to be made per dollar transferred', a switch from project to programme aid for example, 'seems to have the potential for eliminating abundance decisionmaking'.[51] For, 'It changes the lending decision from a microeconomic one to a macroeconomic one . . . Hence the program loan decision has more political weight and visibility and is taken at a higher level in the organization than the decision to finance a project. This type of loan decision, one would think, would lie outside the environment of abundance and not be subject to its perverse economic incentives.'[52]

The 1980s, in fact, saw a switch in the composition of aid along just such lines, in Africa as elsewhere in the Third World. Programme aid, structural adjustment and sector-related lending by the World Bank, for example, rose from a figure of only 7 per cent in the early 1980s, to 27 per cent by the end of the decade.[53] For their part, bilateral aid donors also came to rely more heavily on programme aid, though of course there were variations in the extent of this shift from one country to another.[54]

In many countries of Sub-Saharan Africa, however, project aid to the industrial sector remains extremely important and the need to overcome the biases associated with that form of aid is therefore still a compelling one. One such bias, as noted earlier in this chapter, results from the fact that aid usually finances only the import costs of development projects. A change in this practice, to allow also for the financing of local costs, would accordingly eliminate at least one component of the overall problem.

> Imported equipment, for example, would lose the 'cheapness' it now has, and the tendency to import rather than buy locally would not be encouraged to such an extent. Even though capital extravagance caused by money-moving would continue, the extension of the extravagance to local costs would represent a significant advantage over the current system since it would mean an increase for local industry of the particularly growth-stimulating type of demand that is characteristic of public sector investment programs. . . . The capital extravagance that now takes place, in contrast, hinders the development efforts of local industry in the recipient country. If one is stuck with money-moving decisionmaking, then, the nondiscriminating form of capital extravagance may be less bad than the current discriminating one.
>
> To look upon local costs with the same favor as foreign ones might also give more chance to . . . labor-using [technical] choices.[55]

The technological bias that normally attends 'money-moving' behaviour in donor institutions can be lessened in several other ways as well. A larger number of loan officers, for example, by reducing the amount of aid that each such officer needs to 'move', would bring into consideration technologies that can be efficiently employed at lower output levels.[56] From the aid donor's point of view, however, the appeal of these technologies and the projects that employ them, also depends on the demands they make on costly managerial and administrative resources. It is often argued, for example, that comparatively small-scale projects and technologies are especially demanding in their use of these resources and it is of the utmost importance, therefore, that such demands be reduced to the maximum possible extent. As the donor community itself has come increasingly to recognize, this objective calls, among other things, for a greater degree of reliance on local institutions to implement, manage and monitor investment projects.[57] What this community is also coming to recognize, moreover, is that local institutional capabilities themselves often need to be enhanced in various respects if they are effectively to underpin a movement towards smaller-scale projects and more appropriate technologies.

Ultimately, though, it is the money-moving phenomenon itself that needs to be replaced by an alternative system of rewarding performance in aid donor institutions. In the past, of course, this need has not gone unnoticed by donors, but in recent years the issue appears to have acquired an added and more specific sense of urgency from the World Bank's critical assessment of its own investment projects and from the role therein that inappropriate incentives are thought to have played. A recent Bank review of some 1800 of its projects, for example, found that no less than 37.5 per cent of them were regarded as failures and in some African countries the figure was even higher.[58] Part of the problem, according to the same report, is that loan officers appear to be driven more by lending pressures than by a concern with the successful implementation of development projects.[59]

As a reaction to this, the Bank is reportedly changing its internal structure of incentives: instead of being judged by the amounts of money that they are able to move, managers would be assessed according to how well they 'deliver operations that yield lasting benefits'.[60] It is not at all clear, though, how the Bank leadership intends to address the many problems that will undoubtedly be encountered in making so fundamental a change. (By no means the least important of which problems will involve the separating out of the many other influences that, in addition to staff performance, determine the course and eventual outcomes of investment projects.)

Even in the event that a more result-oriented system of incentives were to be established in the World Bank and other aid donors, however, the technological problems associated with investment projects in the industrial sector would not automatically be resolved thereby. For what essentially needs to be overcome is nothing less than the long and insidious technological legacy of money-moving patterns of decision-making. A sizeable effort would still have to be made, for example, to acquaint decision-makers with technologies that for so long have remained hidden by money-moving behaviour, with its more or less systematic neglect of one end of the technological spectrum and its tendency to 'take', rather than to enhance prevailing technological alternatives. Such an effort, moreover, would need to be directed at a suitably early phase of the project cycle, which, as we have seen, typically lacks a separate technological dimension and which indeed may even divert attention from this aspect of investment projects (as was suggested in Chapter 5). Feasibility studies, for example,

> could be required to: present a detailed costing for several alternative technologies, provide a profile of production techniques in several countries at different stages of development, and indicate why techniques differing from those ultimately recommended were judged unsuitable. Another method to increase the range of technological choice would require parallel efforts by two consulting engineering firms – one from a developed country and one from a developing country. Because the range of knowledge possessed by these firms generally appears to be limited to a few equipment producers in one or two countries, Third World-based consulting engineers are more familiar with machinery produced in developing countries, or less-expensive equipment available from developed countries.[61]

Though they were not in all cases financed by aid donors, the successful cases described in the previous chapter well illustrate the potential that these policy proposals afford in the African context. For, regardless of whether they are privately or publicly owned, those enterprises exhibit a close connection to developing-country sources of technology and technological knowledge, or to sources of second-hand equipment from the developed countries.

Notes

1. See Hirschman's own (1984) discussion of this tendency in his own work.
2. Tendler (1975).
3. Ibid., p. 54.
4. Ibid., p. 85.
5. Renard (1987).
6. See Lall (1992) for evidence of this kind.
7. Tendler (1975), pp. 86–7. John White's study of the impact of external finance on the choice of technology is also worth citing in this respect. In particular, he regards the 'self-portrait of aid agencies' loan officers passively waiting for proposals and constrained by respect for recipients' views' as being 'not one that bears very close examination' (White, 1985, p. 195).
8. Tendler (1975), p. 87. There may also, as White points out, be differences among the aid agencies themselves in the extent to which the principal constraint on increased flows of aid is thought to be a lack of absorptive capacity. In the extreme case, where this view is held very strongly, 'one may end up with project identification missions to stimulate proposals of the kind that the agency wants, project preparation missions to ensure that they are in the form that the agency wants' (White, 1985, pp. 194–5).
9. Gauthier (1990), p. 25.
10. Tendler (1975) cites some interesting case study material from Latin America to illustrate this possibility.
11. Ibid., p. 72. This quotation is applied by Tendler to countries 'less developed' than Brazil. Our suggestion is that it applies particularly well to the African case.
12. Green (1978).
13. Ibid., p. 19.
14. On the basis of detailed plant-level evidence, Kaplinsky (1990) argues that the small-scale variant, using open-pan (OPS) sulphitation technology developed in India, now constitutes the socially optimal choice in many developing countries.
15. Holtham and Hazlewood (1976), p. 147. It is worth noting, however, that Mumias has been the most profitable of the large-scale, state-owned sugar factories in Kenya (Kaplinsky, 1990). In seeking to explain this achievement, Holtham and Hazlewood cite, among several other factors, the technological adaptations at Mumias that reflected the experience of its managing agents, Booker McConnell, in other tropical areas.
16. Ibid., p. 160.
17. Coughlin, Mwau and Begumisa (1988).
18. Kaplinsky (1990), p. 132.
19. Stewart (1979), p. 4.
20. What these projects mostly seem to undergo instead is a financial rate of return analysis.
21. World Bank (1982), pp. 10–31.
22. An aid effectiveness study of four donors in the industry and energy sectors, for example, came to the following conclusion:

    > we found no internationally accepted approach to the choice of technology. Indeed, this appeared to be frequently less well considered than its

importance merits. A number of evaluation reports suggest that avoidable poor choice of technology has led to ineffective projects. *It occurs to us that, by getting their act together so effectively, economists may unwittingly have pre-empted scarce resources, to address their professional concerns, that could have been better spent on examining other facets of projects* (Segal, Quince and Associates, 1984), p. 9, emphasis added.

23. The report cited in the previous note is also worth quoting from this point of view. In particular, the following description is said to describe 'fairly accurately the behaviour of donor agencies in Kenya'.

    Thus, a poor country wishes to increase its output of energy; it makes proposals for hydro-powered generating stations to donor agencies. Some donors will subject the proposal to rigorous cost–benefit analysis. But they may not ask whether a thermal power station or a number of solar-powered units could produce a greater output per unit of input and, if so, what the implications would be for the location of the generating facilities and transmission of the power; and almost certainly will not ask whether the aid resources should go into generating electricity or more traditional forms of energy such as woodfuel (Ibid., p. 9).

24. Leff (1985), p. 68.
25. Cited in Skarstein and Wangwe (1986), p. 153.
26. Ibid.
27. Ibid., p. 153.
28. There is a more general point here that relates to the cases described in Chapter 3 and to other large-scale industrial projects in the public sector. It is that in many such projects, though relatively capital-intensive technologies are justified by the demands of export markets for high product quality, the planned exports fail to materialize, mainly, one suspects, because of an unfavourable macro and trade policy environment. What is in effect an unsuccessful attempt to overcome these unfavourable conditions by product quality improvements contrasts very strikingly with the experience in East Asia, where a favourable policy environment led to the export of labour-intensive products that competed mainly on the basis of price. See Stewart (1987) and James (1987).
29. With regard to exports in particular, the bias towards capital intensity that was ascribed in Chapter 3 partly to the choice of products, may thus at times be more accurately described as a convenient rationalization of 'money-moving' behaviour (as manifest in a preference for large-scale, import and capital-intensive projects).
30. Pack (1986).
31. Holtham and Hazlewood (1976), pp. 213–14.
32. Ibid., pp. 215–16, emphasis added.
33. Lacey (1989), p. 12.
34. Stevens (1981), p. 167.
35. Raphaeli et al. (1984). As Bery (1990) rightly points out, however, the political economy aspects of the Botswana case are as yet far from well understood.
36. World Bank (1994). Just how far the numbers of investment projects have outstripped available managerial capabilities in the poorest African countries

can be gauged from the fact that Burundi's public investment progamme comprises between 500 and 600 projects, while in Tanzania, a very conservative estimate puts the size of the development budget at 2000 projects (ibid). Because these and other investment programmes in Africa typically include a large number of foreign-aid sponsored projects, such estimates also carry implications for what is referred to below as the 'institutional destruction' effect of foreign aid.

37. Ibid, p. 126.
38. Ibid.
39. Morss (1984).
40. Holtham and Hazlewood (1976) make a similar point about the role of technical assistance, which in its concern with improving policy tools, may in fact divert attention from the task of defining the goals of policy.
41. Vitta (1990).
42. Ikiara (1991).
43. Ibid.
44. For a general review of these problems see Vitta (1990) and Unesco (1987).
45. Vitta (1990), p. 1478, citing a study by Goka and Mihyo (1990).
46. Vitta (1990) citing Girvan (1983).
47. Ibid.
48. Wangwe (1986), p. 34. See also Goka, Mihyo and Osunbor (1990).
49. See Ikiara (1991) for a description of the problems confronting the Kenya Industrial Research and Development Institute.
50. Several other studies of the relationship between donors and recipients, in contrast, emphasize the *divergence* of interests between them. Kilby (1993), for example, adopts a principal-agent perspective where the World Bank is the principal and the recipient country government is the agent. According to this view, the Bank has certain objectives which the borrower as agent is asked to promote. Enos (1985) uses a game theoretic framework to explain how conflicting interests between the various parties to aid-financed projects, are ultimately resolved. In particular, he shows how 'overly capital-intensive' technical choices in developing countries can be viewed as the outcome of a three-person game (involving the aid donor, the recipient country government and the owner of a firm in that country) where each player has two strategic alternatives.
51. Tendler (1975), p. 96. This and other donor reforms that would promote more appropriate technology are listed in Stewart (1987).
52. Ibid.
53. OECD (1991).
54. Ibid.
55. Tendler (1975), p. 99.
56. As noted in Chapter 3, the connection between the scale of output and the optimal method of production is often a close one.
57. OECD (1991) contains a detailed discussion of the difficulties involved in channelling aid to small-scale enterprises in developing countries.
58. The report is entitled 'Effective Implementation: key to development impact' and its main results are reported in *The Guardian*, 6 Nov. 1992.
59. The same point is made in *The Economist*, 16 May 1992, which suggests that 'loan officers . . . have been keener to launch new projects than to see them through' (p. 112).

60. *The European*, July 22–5, 1993.
61. Pack (1986), p. 172. This article also contains an extremely useful analysis of the technological impact of World Bank operations in the industrial sector.

References

S.K. Bery (1990). 'Economic Policy Reform in Developing Countries: The Role and Management of Political Factors', *World Development*, vol. 18(8).

P. Coughlin, G. Mwau and G. Begumisa (1988). 'Tied Aid, Industrial Dependence and New Tactics for Negotiations: Observations from Kenya', in P. Coughlin and G. Ikiara (eds), *Industrialization in Kenya: In Search of a Strategy*, Heinemann.

J. Enos (1985). 'A Game-Theoretic Approach to Choice of Technology in Developing Countries', in J. James and S. Watanabe (eds), *Technology, Institutions and Government Policies*, Macmillan.

B. Gauthier (1990). 'Agency Problems and Structural Arrangements within International Economic Organizations: the Case of the World Bank', Political Economy Working Paper 143, School of Business and Center in Political Economy, Washington University, St Louis.

N. Girvan (1983). *Technology Policies for Small Developing Economies: A Study of the Caribbean*, Institute of Social and Economic Research, University of the West Indies, Jamaica.

A. Goka and P. Mihyo (1990). *Technology Policy Institutions in Selected African Countries*, IDRC, Ottawa.

A. Goka, P. Mihyo and O. Osunbor (1990). *Performance Review of Institutions for Technology Policy in Ghana, Nigeria and Tanzania*, IDRC, Manuscript Report 241e.

R. Green (1978). 'The Automated Bakery: A Study of Decision Taking Goals, Processes and Problems in Tanzania', Institute of Development Studies, Sussex University, Discussion Paper no. 141.

A Hirschman (1984). 'A Dissenter's Confession: "The Strategy of Economic Development" Revisited', in G. Meier and D. Seers (eds), *Pioneers in Development*, Oxford University Press.

G. Holtham and A. Hazlewood (1976). *Aid and Inequality in Kenya*, Croom Helm.

G. Ikiara (1991). 'Government Institutions and Kenya's Industrialization', in P. Coughlin and G. Ikiara (eds), *Kenya's Industrialization Dilemma*, Heinemann.

J. James (1987). 'Population and Technical Change in the Manufacturing Sector of Developing Countries', in D. Gale Johnson and R. Lee (eds), *Population Growth and Economic Development: Issues and Evidence*, University of Wisconsin Press.

R. Kaplinsky (1990). *The Economies of Small*, IT Publications.

C. Kilby (1993). 'World Bank-Borrower Relations and Project Supervision', mimeo, Center for Economic Research, Tilburg University, The Netherlands.

R. Lacey (1989). 'The Management of Public Expenditures: An Evolving Bank Approach', Background Paper for the 1988 World Development Report, Policy, Planning and Research, Working Paper no. 46, World Bank.

S. Lall (1992). 'Structural Problems of African Industry', in F. Stewart, S. Lall and

S. Wangwe (eds), *Alternative Development Strategies in SubSaharan Africa*, Macmillan.

N. Leff (1985). 'The Use of Policy-Science Tools in Public-Sector Decision-Making: Social Benefit–Cost Analysis in the World Bank', *Kyklos*, vol. 38.

E. Morss (1984). 'Institutional Destruction Resulting from Donor and Project Proliferation in Sub-Saharan African Countries', *World Development*, vol. 12(4).

OECD (1991). *Development Co-operation*, Development Assistance Committee.

H. Pack (1986). 'The Technological Impact of World Bank Operations', in R. Feinberg (ed.), *Between Two Worlds, the World Bank's Next Decade*, Overseas Development Council.

N. Raphaeli, J. Roumani and A. MacKellar (1984). 'Public Sector Management in Botswana: Lessons in Pragmatism', World Bank Staff Working Paper, no. 709.

R. Renard (1987). 'Aid and Development in Sub-Saharan Africa', University of Antwerp, Centre for Development Studies, Paper no. 87/111.

Segal Quince and Associates (1984). *Aid Effectiveness Study: Review of Four Donors From the Perspective of The Industry and Energy Sectors*, A Contributory Report to Prof. R.H. Cassen's Study for the IBRD/IMF Development Committee's Task Force on Concessional Flows.

R. Skarstein and S. Wangwe (1986). *Industrial Development in Tanzania: Some Critical Issues*, Tanzania Publishing House, Dar-es-Salaam.

M. Stevens (1981). 'Aid Management in Botswana: From One to Many Donors', in C. Harvey (ed.), *Papers on the Economy of Botswana*, Heinemann.

F. Stewart (1979). 'A Note on Project Appraisal and Appropriate Technology – Some Suggestions for Further Work', mimeo, Queen Elizabeth House, Oxford.

F. Stewart (1987). 'Overview and Conclusions', in F. Stewart (ed.), *Macro-Policies for Appropriate Technology in Developing Countries*, Westview.

J. Tendler (1975). *Inside Foreign Aid*, Johns Hopkins University Press.

UNESCO (1987). *Comparative Study on the National Science and Technology Policy-Making Bodies in the Countries of Eastern and Southern Africa*, Science Policy Studies and Documents, no. 66.

P. Vitta (1990). 'Technology Policy in Sub-Saharan Africa: Why the Dream Remains Unfulfilled', *World Development*, vol. 18 (11).

S. Wangwe (1986). 'Technology Imports, Technological Learning and Self-Reliance in Tanzania', World Employment Programme Research, Working Paper no. 165, ILO, Geneva.

J. White (1985). 'External Development Finance and the Choice of Technology', in J. James and S. Watanabe (eds), *Technology, Institutions and Government Policies*, Macmillan.

World Bank (1982). *Technology and Science in World Bank Operations*, Washington, DC.

World Bank (1994). *Adjustment in Africa; Reforms, Results and the Road Ahead*, A World Bank Policy Research Report, Oxford University Press.

Part II
The Political Economy of Bureaucracy

7 The Political Economy of Bureaucracy: A Public Choice Perspective

Only partially addressed in Part I was a question of centrality to our argument which needs to receive separate attention: the importance of bureaucratic objectives. Part II attempts to fill this gap by a political economy approach and in particular one that pays special attention to the role of bureaucracy. We begin by examining the extent to which the public choice approach is able to explain the motivations and behaviour of those who manage state-owned enterprises in the industrial sector. Following a brief discussion of this approach as a whole, we suggest, on the basis of a number of applications, that one particular version is especially well suited to the explanatory task at hand. The concluding section then examines some of the implications for policy that are suggested by this way of viewing technological behaviour in the public sector.

THE NEW POLITICAL ECONOMY

At its most general, the public choice approach, or new political economy as it is sometimes called, 'can be defined as the economic study of nonmarket decision making, or simply the application of economics to political science. The subject matter of public choice is the same as that of political science: the theory of the state, voting rules, voter behavior, party politics, the bureaucracy, and so on. The methodology of public choice is that of economics.'[1] Much of the public-choice literature and especially that part which is devoted to the study of developed-country institutions of the state, falls into the category of what may be called the 'demand-side' of the problem. The preferences of individual voters or members of interest groups, that is to say, are considered responsible for public sector outcomes. According to this view:

> Government, like the market in a pure exchange economy, is viewed simply as an institution for aggregating or balancing individual demands for public policies. Those in government, the candidates and representatives, have been depicted as single-mindedly seeking to be elected. To

> do so they must please voters and interest group members, so that those in government are merely pawns of those outside in a competitive political system.[2]

In that government policy-making is depicted by this version of the public choice approach as being a close reflection of voter and interest group preferences, it is sometimes described as being 'society-centric'.[3] And for just this reason it is regarded by some political scientists as being less applicable in developing countries where 'policy making tends to be more closed, less visible, and more centered in the political executive. . . . Citizens often have their first information about policy when it is formally announced or decreed by the political leadership. In general, high-level administrators and political leaders dominate the policy-making process'.[4] In such circumstances, which in general apply perhaps more readily to Africa than to other regions of the Third World, what is needed is state-centric rather than society-centric explanations, or, in our previous terminology, the supply rather than the demand side.[5]

Though there are differences between the various state-centric or supply-side models of public choice, they have in common the view that government has its own preferences and that these intrude in one way or another on the policy-making process. In the literature on rent-seeking behaviour by government, for example,

> Public officials in developing countries are thought to exchange access to disaggregated public resources – an import license, for example – for some personal benefit, usually economic in nature. In trade theory, this transaction explains the preference for highly disaggregable protectionist measures, the corrupt behavior of public officials, and the difficulty of altering inefficient policy tools.[6]

In the African context, rent-seeking behaviour has recently been used to explain differences in growth rates across a wide range of countries.[7] State-centric ideas have also been used in a well-known attempt by Bates to explain why governments in Africa adopt policies that are blatantly inimical to the interests of most farmers in the region.[8] What has not as yet been applied to the African situation, however, is a micro-economic public choice model that deals specifically with the preferences and behaviour of bureaucrats, who, in Africa as elsewhere in the developing world, make up a sizeable part of the public sector as a whole. It is this strand of the public choice literature, we believe, that helps to explain the behaviour of state-owned enterprises in the industrial sector.

THE POLITICAL ECONOMY OF BUREAUCRACY

Even in relation to the private sector, it has long been argued that in conditions of imperfect competition, firms pursue goals that are related to the size rather than the profitability of the enterprise. In a well-known early contribution to the theory of firms, Baumol, for example, argued that in such circumstances managers seek to maximize sales, or, in a more realistic dynamic context, to maximize the rate of growth of sales. 'Economists who have spent time observing the operations of business enterprises', he claimed, invariably 'come away impressed with the extent of management's preoccupation with growth. Expansion is a theme that . . . is dinned into the ears of stockholders and is constantly reported. . . . Indeed, in talking to business executives one may easily come to believe that the growth of the firm is the main preoccupation of top management'.[9] Similarly, according to Oliver Williamson's equally well-known theory of managerial discretion, salaries, perquisites, staff-size and control over investment funds are the overriding concerns of management and each of these variables is likely to be promoted by an increase in the size of the firm.[10]

Institutional size is also central to the efforts that public choice theorists have made to explain the behaviour of institutions in the public rather than the private sector.

The first systematic attempt to study bureaucracies from a public choice perspective was undertaken by William Niskanen, whose specific concern was with

> organizations that have *both* of the following characteristics:
> 1. The owners and employees do not appropriate any part of the difference between revenues and costs as personal income.
> 2. Some part of the recurring revenues derive from other than the sale of output at a per unit rate.
>
> In a single sentence: *Bureaus are non-profit organizations that are financed at least in part, from a periodic appropriation or grant.*[11]

Thus deprived of an incentive to maximize profits, like his counterpart in the Baumol or Williamson theory of the firm, the manager of a bureau in Niskanen's model turns instead to many of the same objectives that have just been mentioned in relation to the privately owned firm, such as 'salary, perquisites of the office, public reputation, power, patronage, output of the bureau'.[12] Since these objectives are all assumed to vary directly with the size of the budget, the aim of the bureaucrat thus becomes one of maximizing this variable during his tenure in office. (The time issue is important partly because most of the gains that accrue to the budget-maximizing bureaucrat 'are nearly unrelated to the "net worth" of his organization after his departure' and partly because his tenure itself may in some country circumstances be of relatively short duration.) In his pursuit of this goal, however, the budget-maximizing manager is said to be motivated not only by his own self-regarding quest for utility maximization, but also by other actors, whose preferences according to Niskanen, exert pressure in the same direction.

What he has in mind here are not only the demands for larger budgets from subordinates within the bureau itself (to the extent that larger budgets are thought to enhance the prospects for job promotion, job security, etc.), but also and less obviously, the demands of the same kind from the very institutions that exert control over the bureaucrat's tenure, in that, for example, 'both the executive and legislative officers reviewing the bureau fully expect the bureaucrat to propose aggressively more activities and higher budgets. Indeed, they would not otherwise know how to perform their review role. They depend on the bureau to seek out and propose these and to make a case for larger expenditures in old programmes.'[13] Exerted as they are by two groups of persons with a significant say over a bureaucrat's tenure in office, these influences form part of what Niskanen calls the 'will to survive' basis for budget maximization and they reinforce the 'rationality' (or utility maximization) argument in favour of this goal.

The mechanisms through which budget maximization is thought to promote the goals described above in connection with the rationality argument, are neither difficult to identify, nor particularly hard to believe.[14] It is eminently plausible, for example, to imagine that larger budgets may often make it easier for bureaucrats to divert resources to the creation of 'perks' (such as bigger or more exclusive offices) and to negotiate higher salaries on the grounds that an enlarged bureau carries more responsibility. Bureaucrats in larger agencies 'control more patronage and have a greater capacity to influence events and overcome resistance. They can also create organizational slack and resources more easily to meet unusual risks or crises when they occur'.[15] Or, through still another mechanism, larger size tends to be associated with 'enhanced public reputations for influence, and higher status among their peer group. Senior civil servants commonly regard the comparative budget increases across departments as indicating winners and losers in the . . . competition for resources'.[16] These and other mechanisms are presented more systematically in Table 7.1 which shows how each of the bureaucratic objectives mentioned above are promoted by budget maximization. To some extent, of course, these goals are interrelated

Table 7.1 How bureaucratic objectives are served by budget maximization

Power	*Salaries and perquisites*	*Patronage*	*Output*	*Reputation*
Budget maximization may lead to an increase in economic power (via monopoly or oligopoly) and in political power via an enhanced influence over other state institutions (for example those that make and implement economic policies) and a greater capacity to resist controls imposed by other institutions, especially when large size is accompanied by greater complexity of the enterprise	Budget maximization simplifies the task of diverting resources into perquisites; it facilitates the payment of salary increases (e.g. because of the extra responsibility that large size may entail) and job prospects are enhanced in rapidly growing enterprises	Budget maximization offers scope for patronage in terms of functions (e.g. jobs, board members, contracts for input supplies) and in terms of regions that are politically favoured (and that benefit, for example from the provision of a certain service, such as electricity)	Large budgets lead to an increase in output, depending on how much of the budget is invested (rather than consumed), and on the size of the incremental capital–output ratio (assuming that some part of the budget is invested)	'Proving managerial success with reference to output or revenue data is of particular interest for public enterprises. The latter are often prevented from seeking maximum profits and therefore their managers' success cannot be appraised with reference to profit data. An interesting example of the practical application of *output maximization* by a public enterprise is the maximization of passenger miles pursued by London Transport some years ago.'[1] Another example is the way in which development finance institutions promote their achievements in terms of increases in their size (see Table 3.2a)

[1] D. Boes (1989), p. 297; emphasis added.

and the pursuit of one goal then also has implications for another (as when, for example, output increases also have implications for the reputation of an enterprise or when an improvement in the latter also enhances its political power).

The foregoing discussion of the budget-maximizing bureaucrat does not of course purport to summarize more than a part of what Niskanen has to say about bureaucracy. He is also especially interested, for example, in the implications of the non-market nature of the output supplied by a government bureau. Indeed, 'a bureau does not typically supply a number of units of output as such, but levels of *activities* from which output levels must be inferred. Thus the Department of Defense maintains numbers of combat personnel and weapon systems, although it supplies various degrees (units) of defensive and offensive capabilities.'[17] This particular aspect of Niskanen's approach need not detain us, however, for the questions we have posed mainly involve public enterprises rather than bureaux and the former, unlike the latter, do engage in the production of marketed output and they do, at least to some extent, rely on the receipts therefrom to finance their activities.

THE POLITICAL ECONOMY OF BUREAUCRACY IN AFRICA

As we have just described it, what the new political economy has in common with our earlier analysis of the bureaucratic choice of techniques is the primacy it accords to the role of finance in the decision-making process (though of course our treatment emphasized foreign exchange rather than the budget in general). Niskanen's account, however, also extends our discussion in several key respects, the first of them being the additional insight it offers into the relationship between the time element of production on the one hand and the factor intensity of that production on the other.

I am referring here specifically to what Niskanen (building on earlier work by de Alessi)[18] refers to as the 'overcapitalization hypothesis'. The idea is that the 'overcapitalized' nature of production techniques in the public (relative to the private) sector can be ascribed to the relative brevity of the time period over which the bureaucrat is assumed to pursue his own objectives. In particular, since the 'bureaucrat's rewards are specific to his tenure in that position' he is led thereby to 'prefer present spending to future spending and, thus, to prefer production processes with higher capital costs and lower operating costs'.[19] This hypothesis is supported, according to Niskanen, by studies showing that public utilities are more capital intensive than private utilities in the electricity sector. It is also supported by the results of the comparative studies mentioned in Table 1.6 (somewhat inconclusive though those were shown to be).

The second and more important respect in which the public choice approach extends our previous analysis has to do with a certain degree of complementarity between the two. Our discussion in Part I, it will be recalled, focused largely on the question of how new projects (typically embodying external finance of one kind or another), are used by managers of state-owned enterprises as a means of maximizing the supply of this type of finance under their control. Niskanen's analysis, on the other hand, is more concerned with the ways in which budget maximization promotes a particular set of goals that are allegedly held by the bureaucracy (though when it is applied to the bureaucratic choice of techniques his analysis comes to a conclusion not dissimilar from one of our own, in Jackson's words, that: 'When undertaking the appraisal of a new project, it would be in the interests of the budget maximizing bureaucrat to understate the costs and to overstate the benefits of the project and at the same time to use a low rate of discount when calculating expected net benefits to the community. This kind of strategy increases the chance of the project being approved and the continued existence of the bureau.')[20]

Expressed in terms of their contributions to a causal explanatory sequence, one can say that the Niskanen analysis takes off from where the bureaucratic choice of technique ends. Thus, to the extent that his model – originally formulated in relation to the bureaucracy in the United States – is applicable to the situation in Africa, it is capable of providing exactly what is missing from our earlier discussion: namely, an analysis of how foreign exchange (and more generally budget) maximization contributes to the objectives that managers of state-owned enterprises really pursue. Let us accordingly assess the Niskanen model from this point of view and we shall find it convenient to order the discussion around the bureaucratic objectives that he himself posited.

POWER AND ITS EXPLOITATION

It was apparent from Tables 1.1, 1.2 and 3.2 that public enterprises in the industrial sector of numerous African countries grew very rapidly in the 1960s and 1970s. These rapid growth rates were partly due to the creation of entirely new public enterprises and partly to the expansion of existing firms. (In Tanzania, for example, where growth rates of parastatals were rapid even by African standards, the value of investments in new companies exceeded both the value of investments in existing parastatals and in the amounts that were devoted to nationalization).[21] From either basis, however, the result was a marked increase in the numbers employed by public enterprises in the (formal) industrial sector. In Ethiopia, for example, the growth rate was such that by 1982 86,000 people were employed in the manufacturing sector alone.[22] In Ghana, by that same year, public enterprises in the industrial sector employed almost 60,000 workers.[23]

In degrees that are likely to vary from one country to another, these growth rates should be viewed as both a growing source of political power for the state-owned industrial sector, and a reflection of deliberate efforts on the part of that sector to increase its hegemony over other participants in the political economy of post-colonial Africa. According to the former view of how the causality runs, influence accrues over time to the expanded numbers of parastatals in much the same way that the (no less rapid) growth in the civil service led 'almost inevitably' to the creation of 'a privileged group that gradually developed corporate interests of its own. . . . Civil servants and government employees . . . emerged as the core of a new *dominant class in the postcolonial period*'.[24] Thus, although individual public enterprises may often have been created for other purposes – such as to replace a weak private sector, to promote equity, or to secure the commanding heights of the economy – the sheer extent of the phenomenon meant that *in the aggregate* the sector as a whole became a distinct political force to be reckoned with.[25]

The latter view of how the growth and the extent of influence of public enterprises are related, in contrast, regards these institutions essentially as instruments in the pursuit of power; as such, therefore, the growth of public enterprises becomes a direct consequence of a prior propensity to seek more power rather than a cause of any such increase. The particular appeal of the public enterprise from this point of view is that it affords a relatively high degree of autonomy from other state institutions and a correspondingly sizeable degree of managerial discretion over operating procedures, in comparison, say, with the relatively rigid regulations governing hiring, pay, budgeting, etc. in the civil service.[26] In comparison with bureaux, for example, the fact that the output of a public enterprise is sold means that its performance can more readily be evaluated. As a result, there is less need to resort to the highly restrictive regulations that are usually applied to other public sector institutions. It is precisely this kind of discretion, it seems, that would be so valuable to the power-seeking manager in the public sector, who presumably would also go to some lengths to protect so valued an asset if it came under threat in some way.

Regardless of whether it is acquired by one or another (or even some combination) of these mechanisms, however, such power as does accrue to public enterprises in Africa often seems to be employed as a means of self-enrichment rather than as (in Weber's theoretical ideal) 'a juggernaut which, because it was technically competent but politically neutral, would necessarily be controlled from the top downwards'.[27] This tendency is frequently described in the political science literature: David Leonard, for example, claims that 'state organizations in Africa are extensively used to pursue informal, personal goals of their managers rather than the collective ones that are formally proclaimed'.[28]

The literature on the topic, however, is far from being purely academic. In a range of African countries, commissions appointed by governments themselves have documented the variety of ways in which the power afforded to managers of state-owned enterprises has been exercised as a means of self-enrichment, often it seems, with the collaboration of foreign suppliers, managers and financiers.[29] Of the numerous explanations commonly advanced in the political science literature to explain this seemingly ubiquitous tendency, the following are perhaps not untypical. 'After all' writes Robert Williams:

> The new states were . . . European creations based on arbitrary and artificial boundaries. Thus, it could not be assumed that the inhabitants would be united by any sense of mutual interest, shared history or common destiny. In practice, status and influence were locally acquired and enjoyed and, as a consequence, the locality rather than the nation was the normal focus of loyalty and support. The absence of any widely shared concepts of citizenship, brotherhood or national interest implied that people of different ethnic groups, religions, regions or localities were more vulnerable to exploitation. To most Africans, the nation was a vague and meaningless abstraction and government was a remote and alien tyranny whose resources were a legitimate target for diversion and redistribution.[30]

SALARIES AND PERQUISITES

Measured in terms of their percentage contribution to total government expenditure, wage and salary payments to public sector employees in Africa are strikingly high in comparison with all other regions. This is evident from Table 7.2 which shows that in 1979 the percentage in question was

Table 7.2 Current expenditure on wages and salaries as a percentage of total government expenditure and lending minus repayments

	1972	*1974*	*1976*	*1978*	*1979*	*1980*
World	18.3	16.6	15.3	14.8	14.7	n.a.
Industrial (market) countries	16.6	15.5	14.0	13.3	13.0	12.3
Oil-exporting countries[a]	31.2	18.7	19.0	23.1	30.4	n.a.
Africa	35.2	30.9	28.5	28.8	29.2	29.7
Asia	n.a.	18.9	17.4	15.8	15.3	14.7
Middle East	n.a.	n.a.	14.5	16.4	15.0	n.a.
Western Hemisphere[b]	22.1	20.6	21.3	21.1	20.8	n.a.

[a] Africa's most populous country, Nigeria, is included under the 'oil-exporting' and not the African heading.
[b] 'Western Hemisphere' refers to the developing countries of Latin America and the Caribbean.
Source: Abernethy (1988), p. 191.

Table 7.3 Non-financial public enterprises' share of non-agricultural employment

	OECD countries	*Developing countries*				
		Total	*Africa*	*Asia*	*Latin America*	*N. Africa*
Average share (%)	4.1	13.9	18.7	15.7	5.5	10.3
Number of countries	14	18	8	4	5	1

Source: Heller and Tait (1984), Table 4.

almost double that for Asia and the world as a whole and about 50 per cent higher than that for Latin America and the Caribbean.

Part of what makes Africa distinctive from this point of view is that in this region the public sector in general and public enterprises in particular (see Table 7.3), are comparatively more important sources of employment than elsewhere.

Another part of the explanation, however, has to concern itself with public sector wage and salary differentials between Africa and other regions, since these may also have influenced the overall differentials described in Table 7.2. Although it includes for comparison only the Asia region and though it is not based on a large sample of countries, the

Table 7.4 Earnings as a multiple of per capita GDP: Sub-Saharan Africa and Asia

	Sub-Saharan Africa			Asia
		Range		
Earnings	*average*	*low*	*high*	*(average)*
Central government	6.9	3.9	15.1	3.3
Manufacturing	4.3	1.7	6.5	2.8
Construction	3.1	2.0	4.1	2.0

Note: Data cover 13 IDA-eligible countries in Sub-Saharan Africa and three low-income countries in Asia for various years from 1978 to 1983.
Source: World Bank (1986), p. 21.

estimates in Table 7.4 lead one to suspect that from this point of view as well, there may be something distinctive about the African case. For, in all three of the public sector categories mentioned in Table 7.4, the average estimate of earnings as a multiple of per capita GDP is substantially higher for Africa than for Asia. Thus, to the extent that the Asian experience reflects the situation in the Third World as a whole, the African bureaucracy appears to be not only extensive but also well paid in relation to its counterparts in other developing countries (see below, however, regarding the decline in real wages during the period of structural adjustment reforms). And, as with so many other distinctive characteristics of the African state, an adequate explanation of the earnings differentials shown in Table 7.4 needs to incorporate the socio-political influences that were so powerfully exerted during the colonial period.

According to Abernethy, for example, 'One must examine on the one hand the character and policies of the governments established by European colonial rulers, and on the other hand the initiatives taken by Africans as they coped with the condition of being ruled by racially and culturally distinctive foreigners.'[31] With regard to the former question, the dominant influence on what was to become the 'conspicuously consumptive African state', took the form, he suggests, of salaries and service conditions that were generous enough to attract 'well-educated, competent Europeans into the colonial administrative service'.[32] Thus it was that European levels became the frame of reference for salaries and working conditions in the African civil service, during and indeed throughout the entire colonial period.

African responses to these externally determined salary levels – the second element of the approach suggested by Abernethy – varied according to the socio-economic status of particular groups in the population. At the bottom were peasants, who, by virtue of their relatively low incomes and demanding mode of work were intent on gaining access to low levels of the bureaucracy. For them, 'The bureaucrat, by virtue of the resources he commanded and his links to the larger structures of government, exercised – or at least symbolized – power, wealth, status, and security.'[33]

For bureaucrats already at the lower levels of the civil service hierarchy, the relevant frame of reference became the Europeans who occupied the upper echelons of the civil service 'whose income and perquisites might be worth, in the extreme case, over one hundred times the material rewards at the bottom'.[34] Initially the aspirations to which this kind of comparison gave rise could scarcely be met by Africans whose educational attainments lagged so markedly behind European levels. Over time, however, as educational achievements grew, the dominance of Europeans in the upper strata of the bureaucracy was increasingly called into question and the force of this challenge was made all the more powerful by the emergence in Africa of nationalist movements of one type or another.

At one level, nationalism took the form of a collective desire for: '*equality with Europeans* in access to top posts and in remuneration for a given level of responsibility. That nationalists focused so much of their attention on this matter is hardly surprising. It was an issue on which all Africans, despite their differences and rivalries in other respects, could agree. It was an issue that touched a deep emotional chord.'[35] Because of the prospect that it raised for widespread advancement in a reformed bureaucracy, nationalism also served more narrowly to promote the 'personal career interests' of leaders whose political fortunes were closely allied to government workers. Far, therefore, from eschewing the salary scales that the European colonialists had established, nationalist leaders in Africa,

> firmly *insisted* upon retention of these scales. Ironically, retention was justified on egalitarian grounds, even though its effect was to institutionalize within the independent polity the powerful inegalitarian legacy of European rule. The reference group for the nationalist leaders was still the colonialists who were departing, not the peasants and urban poor who remained.[36]

The bureaucracy that emerged in Africa during the post-colonial years, then, was one whose principal features conformed rather closely to what one would have expected on the basis of a public choice approach to the topic. And again in conformity with that approach, the emergent African bureaucracy was to prove strikingly resilient in the face of the numerous challenges to its dominance that were subsequently mounted. To quote Abernethy again:

> Once the expensive state is in place, political leaders find it extremely difficult . . . to institute reforms reducing the size or pay scales of the bureaucracy. The national bureaucracy can readily respond to such a challenge by transforming itself into a trade union, practicing go slow tactics, or going out on strike. In theory, bureaucrats are the agents of government output. In practice, the larger and more privileged their ranks relative to the population they govern, the more they should be analyzed as powerful *input* agencies within a political system. The salary structure of African governments has helped transform these governments into powerful interest groups whose overriding interest is themselves. With the system reproducing itself in this manner, able young people face continuing strong incentives to eschew private sector entrepreneurial activity in favor of government employment. The prospects are thus reduced that an indigenous and autonomous bourgeoisie . . . will emerge.[37]

During the 1980s, however, under strong pressure from the World Bank and the IMF to reduce their burgeoning budget deficits, many countries in Sub-Saharan Africa witnessed a drastic decline in the real wages of those employed in the public sector, although to some extent these reductions may have been offset by increases in perquisites of one kind or another, and though few states seemed able to combine these reductions with a marked diminution in the numbers employed in the public sector.

PATRONAGE

The two previous sections have suggested that certain distinctive features of the African bureaucracy make it especially amenable to a public choice type of analysis. The same argument can be made in relation to patronage, an objective that is emphasized by both Niskanen and those political scientists who have studied the African state. Indeed, of the numerous terms in which the bureaucracy has been described by this political science literature, few have received as much attention as the obligations imposed on the holder of high public office by his various clients. Typically, the prevalence of these patron–client relationships is explained along much the same historical lines, irrespective of whether the author's particular account happens to be framed in terms of 'personal rule', 'neopatrimonialism', 'the economy of affection', or some other similar concept of the state.[38] In brief, what the typical explanation seems to amount to is as follows.[39]

The circumstances in which the post-colonial state emerged in Africa were strikingly different from those that prevailed during the early phases of capitalist development in the now developed countries (or, for that matter, from the circumstances in the contemporary experience of certain developing countries as well). In Western Europe and Japan, so the argument goes, the modern state emerged at a time when class differences were already apparent and a particular class had become dominant. Moreover, the social and cultural circumstances in these countries were then generally such that the institutions of the state were able to command relatively wide support from the society at large. 'In sum', writes Sandbrook, 'the Western state today is disciplined and responsive because of the demands of class power and a pluralist political system.'[40]

In both these respects, however, the state that emerged in post-colonial Africa had a very different and historically distinctive character. For one thing, instead of forming a society that was strongly divided along class lines, the post-colonial African state was founded principally on the basis of communal ties of a pre-capitalist kind, involving, as they often do, an extended network of social obligations (such as ties to the extended family, to the local community and to specific ethnic groups).[41] Nor was the emergent state able to command anything like the same degree of societal support that was exhibited in the historical experience of the West European countries.[42] In part this was a reflection of how the state had been perceived during the colonial period – typically as a somewhat remote and alien force. But it probably also reflected the comparative heterogeneity and – from the point of view of a national social and political identity – divisiveness of pre-capitalist African societies. It is sometimes said for example that the organization of social and economic life in those societies 'imposes social obligations on individuals that limit their interest and capacity to support public concerns outside their community, however defined. It tends to perpetuate a locale-specific outlook which often comes into conflict with the principles on which national development is based.'[43]

Thus it was that a state with communal, rather than class or populist, ties came into being in post-colonial Africa. And in this 'patrimonial' state it is not at all surprising that development projects should come to play so crucial a political role, for, unlike price policy instruments, projects bestow on those who run them the ability to grant selective benefits of one kind or another on a wide range of potential claimants.[44] The scope thus afforded for political patronage, moreover, will tend to increase with both the number and the size of development projects.

If this consideration helps to explain the appeal of projects in the public sector, it is also helpful in explaining why such projects have so often performed poorly from an economic and financial point of view. For, as Bates has put it:

> public officials are frequently less concerned with using public resources in a way that is economically efficient than they are with using them in a way that is politically expedient. If a project fails to generate an adequate return on the public investment but nonetheless is privately rewarding for those who build it, provision it, staff it, or hold tenancies in it, then political officials may support it, for the project will serve as a source of rewards for their followers.[45]

Though Bates bases this conclusion mainly on his study of agriculture in Sub-Saharan Africa, it would seem just as relevant to industrial projects in that region. And within industry, it applies to the technological no less than to other determinants of the efficiency with which resources are used in public sector projects.

OUTPUT GROWTH

Output growth, we suggest, is pursued less as an objective in its own right and more as a means of promoting the other goals that have just been described.

It has already been noted, in relation to Table 7.1 for example, that managerial success in the public sector may often be measured in terms of output (or sales) growth rather than in terms of profitability. And we have also argued, in Chapter 3, that during the time when import substitution was regarded as the preferred development strategy by most African countries, the reduction or elimination of excess demand for important wage goods carried particular weight in the public sector. As such, therefore, rapid output increases of these goods are likely to have conferred considerable status on the management of a state-owned enterprise even when – as was often the case – the output growth was achieved at considerable resource-cost.[46] Such reputational effects, in turn, may have made it easier for public enterprises to justify not only wage and salary increases but also additions to the numbers of workers that they employed.

CASE-STUDIES

While there is a large political science literature on which one can draw to make a rather general case for a public choice interpretation of the role of bureaucracy in African industry, there is much less evidence at the level of specific institutions in the public sector. This is partly a reflection of the fact that most of the literature at the institutional level is microeconomic in character, and as such tends to concern itself with the (economic) consequences rather than the causes of particular forms of behaviour. Such explanations as are normally offered for this behaviour, moreover, tend to be too restrictive in their scope to throw much light on the political variables that a public choice oriented framework would entail. Even the relatively few case-studies that are available in the political science literature, are not written specifically from a public choice perspective. Nevertheless as the following section tries to show, some of these case-studies usefully illustrate much of what is most central to that perspective.

Public Enterprise and the Energy Sector in Nigeria

From the point of view of our earlier discussion, one of the most interesting aspects of Wilson's study of rural electrification in Nigeria's Mid-Western state in the early 1970s,[47] has to do with the institution that was chosen to organize the programme. For, while there were at least two other organizational forms that might have been selected, senior officials in state government elected instead to create an entirely new public enterprise. And the reason for this choice, according to Wilson, 'was as much to remove the activities of electricity supply and distribution from the somewhat tighter bureaucratic controls of the Ministry of Works so that they could be perverted for personal use as it was to ensure efficient operation of the program.'[48] Economic rationality within the public sector, that is to say, was sharply at odds with the type of political rationality that is posited by public choice theory.

In particular, there were at least the following five respects in which the newly created public enterprise was made to serve the private interests of those who were charged with managing the programme:

1. To gain privileged access to services produced by the PC [public corporation] (its outputs);

2. To gain privileged access to the supply inputs required by the PC (the foremost being capital, but including machinery, labour, etc.);
3. To gain privileged access to the receipts and revenues generated by the PC;
4. To engage in kickbacks, double billing and other practices for expensive foreign equipment;
5. To gain access to potential political power through the authority to decide what individuals, groups or firms will or will not receive a 'public service' such as electricity. This bargaining power can then be translated into other spheres of political and economic activities.[49]

The time element too seems to have been important to the key decision-makers, not least because of the influence it exerted on the choice of technology for the electrification programme. Two distinct technological options presented themselves: one was based on thermal electricity generators capable of installation anywhere in the state and which ran on diesel fuel, while the other alternative was to link the rural locations to the existing national electricity grid. The latter it seems was not only less expensive but also technically less demanding (in that thermal installations need to be kept under constant and careful surveillance and to be kept fuelled). Nevertheless the latter alternative was chosen, partly on the grounds that it enabled electricity to be brought more rapidly to the areas in question. This 'time factor' writes Wilson 'was politically relevant to the elite, and it overcame objections raised on technical grounds'.[50] From a financial point of view as well, the time element favoured the thermal option, for if the national grid transmission had been selected, the public enterprise 'would have had to disburse the money at a slower rate over a longer period of time. This was a critical issue . . . and there was pressure to disburse the funds quickly.'[51] (So rapidly, in fact, were the funds disbursed to meet excessive equipment and other costs that the entire budget was exceeded well within the plan period as a whole.)

Precisely on account of the opportunity for self-enrichment that it so clearly provided in this case, the public enterprise proved highly attractive to a bureaucracy that in name at least was bent on meeting society's growing needs for energy. Indeed, because of the financial and organizational autonomy that it afforded to its managers and promoters, ownership by the state was chosen from among other institutional possibilities in eight out of the other twelve states in Nigeria as well.

PARASTATALS IN TANZANIA WITH PARTICULAR REFERENCE TO THE NATIONAL DEVELOPMENT CORPORATION

In the Hirschman tradition, this book set itself the task of explaining a number of seemingly paradoxical forms of economic behaviour in the public sector. As will be recalled from the Introduction, all these paradoxes relate in one way or another to the case of Tanzania and most of the subsequent chapters have, accordingly, paid particular attention to that country's experience. If, as a result, we have made a certain amount of progress towards discovering the rationality that underlies the observed behaviour of public enterprises in Tanzania, our analytic framework nevertheless remains an incomplete one. What it lacks above all is an understanding of the political context in which this behaviour took place.

According to what form of political logic, for example, were managers so preoccupied with variables that govern the extent as opposed to the profitability of their operations? For what political reason was (and for the most part apparently still is) the initiation of new projects so much more important to most of those enterprises, than the management of and the learning from existing projects? What were the political relationships between parastatals and various other branches of the state and what effects did these have on the behaviour of the former? Fortunately, the political science literature contains an excellent study by Mukandala of the period before and after the Arusha Declaration,[52] which enables these questions to be addressed not only at the level of a particular parastatal but also at the level of the public sector as a whole.

Briefly stated, Mukandala's thesis is that the growth and behaviour of parastatals in Tanzania form part of and need to be understood in relation to a protracted and intense conflict between the 'political stratum' of the state on the one hand and the 'managerial stratum' on the other (a distinction that corresponds roughly to the making of policy as against its implementation by appointed agents of the state). Though parastatals had of course existed in Tanzania prior to the Arusha Declaration, it was not until the 'emergence of the parastatal sector into a strategic and complex branch of the state'[53] in the years immediately thereafter, that the political and managerial strata came into direct and open conflict. For by then parastatals had become 'too big, diverse and strategic to be left to the managerial stratum yet proved too difficult to control through conventional government methods'.[54] It was already apparent, for example, that far from fulfilling their assigned role as vehicles in the transition to socialism, the parastatals were in reality behaving more like 'bastions of capitalism', with very little connection to the political institutions of the state. Of particular concern to the 'political stratum' was the National Development Corporation (NDC) which accounted for more than 50 per cent of investment by parastatals in new firms and which controlled most of the state's assets in manufacturing and other sectors of the economy.

Thus, beginning in 1969, the political stratum embarked on a series of measures that were intended to curtail the degree of freedom then being enjoyed by the parastatal sector in general and the NDC in particular. This marked the start of what, for reasons that will shortly become apparent, Mukandala refers to as the 'rationalization' phase in the establishment of parastatals in Tanzania.

One such measure for example saw the creation of parent ministries in whose control particular groups of parastatals (in say agriculture or industry) were placed. Other measures involved a heavier reliance on certain parastatals (such as the National Price Commission and the Bank of Tanzania) to regulate the behaviour of other parastatals, and a strengthening of the role played by central ministries (such as the Treasury and the Ministry of Development Planning). The most drastic of all the weapons employed by the 'political stratum' after 1969, however, was the attempted 'rationalization' of the parastatal sector. What this essentially meant was the splitting up of then-existing corporations into progressively smaller units in the hope that they would thereby become more manageable. Over the period from 1971 to 1974, for example, the NDC lost no fewer than 17 of its operating companies and 19 of its projects as part of the drive towards 'rationalization' of the public sector.

On the whole, however, these and other attempts to rein-in the rapidly growing influence of the 'managerial stratum', met with remarkably little success. In part this was a reflection of a rather predictable range of administrative and logistical problems. The sheer extent of the parastatal sector, for example, made life extremely difficult for those who were supposed to exercise control over it. Then too there were the problems caused by the lack of technical expertise on the part of the controlling institutions. To some extent the controls themselves were poorly conceived, as was the case for instance with areas of responsibility that were left 'largely undefined' or with the vagueness of the objectives assigned to the parastatal sector. To a considerable degree, however, the failure of the 'political stratum' to impose its will on parastatals in general and the NDC in particular was due to the vigorous counter-measures that this sector *itself* undertook.

It is striking how closely some of those measures conformed to what one would have expected on the basis of the public choice analysis of bureaucracy described earlier in this chapter. Consider first from this point of view Mukandala's account of how the parastatal sector as a whole responded to what, in the name of rationalization, was effectively a standard 'divide and rule' tactic on the part of the 'political stratum':

> On the one hand the managerial stratum established new subsidiary companies to replace those hived-off by the state . . . This was deemed necessary because the more subsidiaries a corporation had, the bigger it was and consequently a) more resources were allocated to it in the national budget for developmental purposes; b) the more bargaining power it had when negotiating for higher salaries (especially for its management). Size showed 'umuhimu' or importance relevance; c) strengthened its hand in mobilizing external finance on its own. 'Going international' was useful in two key ways. First it meant unrestricted foreign travel for the managerial stratum and the consequent allowances and a chance to purchase foreign goods (especially in times of shortages 1978–1985). Foreign travel became such a big thing that in late 70's a new passport type called a 'service passport' was issued to make travel for the various petit bourgeoisie strata easier. Second, and most importantly, external finance was also useful because once obtained, it assured approval for the projects from the Treasury, the Bank of Tanzania and the Ministry of Development Planning. Foreign capital's need to invest in new projects rather than expanding or rehabilitating old ones reinforced this trend.[55]

From the point of view of a sector engaged in a struggle for its very political survival, therefore, what seemed to matter most was its *size* and *power* rather than its efficiency and profitability. And no parastatal in Tanzania engaged in a more intense contest over its size than the National Development Corporation. As the largest holding company in Tanzania at the end of the 1960s, the NDC was not surprisingly an especially inviting target of the efforts on the part of the 'political stratum' to regain control over the parastatal sector as a whole. The most important of those efforts was, as already noted, the fragmentation of the corporation into progressively smaller units. By the end of 1974 this process had already resulted in the loss of 17 companies and following the adoption of a 'basic industry strategy' in 1976, further losses ensued. In particular,

> Between 1977 and 1979, NDC . . . lost thirteen companies including the prized beer and cigarette companies, as well as the leather, paper and printing industries. The following year, NDC lost a further 17 companies and projects. The corporation lost 21.9 per cent, 53.3 per cent

and 31.5 per cent of its total investments in 1977, 1979 and 1980 respectively.... By the end of this phase, NDC's domain was confined to metals and metalworking industries, mechanical engineering industries, electronics and electrical engineering primary metals, industrial gases, and pulp and paper.[56]

The NDC responded to this challenge in much the same way as other parastatals: it engaged in the 'hatching' of subsidiaries. 'The more the state hived-off, the more subsidiaries NDC started or took on. The corporation's 1973–75 three year plan, for example, was premised on the injection of Shs 140 million by the stratum into the corporation for expansion and new projects partly through the take-over of private concerns.'[57]

There was more at stake, however, than just the creation of increasing numbers of subsidiaries. For by having to devote so much time to its investment function, as a parastatal holding company the NDC was in effect neglecting its other, managerial, function; that is to say its role as manager of the subsidiaries that such investments helped to create. More specifically,

> Initiating new projects, mobilizing finance from internal and domestic sources and setting up new projects as subsidiary companies all expanded the size of the corporation, increased its status, and helped to defend its budget share. Managing the subsidiaries was a more difficult task, having little glamor. *The apparent managerial unresponsiveness to pressure from the political stratum to improve company effectiveness and efficiency stems in part from this structural imperative.*[58]

It is this same imperative, we suggest, that lies at the core of the technological behaviour exhibited by much of the parastatal sector in Tanzania. On precisely the same grounds, one can argue, for example, that the efficiency of a particular technology that happened to be associated with an investment project would not have entered the political calculus of a parastatal holding company in any important way. Nor would other technological characteristics, such as labour-intensity, self-reliance and decentralization, which were important to the political stratum. From the point of view of parastatal managers, there 'simply was no time for these'[59] during the 'frantic spate of subsidiary creation' (which led in fact to 'an increase rather than a decrease in the use of foreign consultants, technical experts and management agents. The national call for localization only took effect at the headquarters level. Also, there was a renewed aggressiveness toward recruiting foreign project partners, the call for self-reliance notwithstanding. . . . This in turn led to a faster increase in the importation of capital goods rather than intermediate goods and over-installation of new industrial capacity'.[60])

The effect of all of this on technological capabilities is not difficult to imagine. On the one hand, the political pressure to create subsidiaries meant that insufficient time was available to the parastatal holding company for undertaking the various types of effort on which the acquisition of technological capabilities so crucially depends. A management agreement with a foreign partner, for example, was more likely to have been viewed by the parastatal holding company as a means of overcoming the immediate skills constraints on its expansion, than as an issue with important long-run implications for the acquisition of indigenous technological capabilities.[61] On the other hand, in neglecting its managerial function the parastatal was effectively eschewing a whole range of issues that bear on technological learning and mastery at the operational level (and that have to do for example with project execution, repair and maintenance).[62]

STRUCTURAL ADJUSTMENT AND THE POLITICAL ECONOMY OF PARASTATALS IN TANZANIA

By seeking to diminish the size of the public sector and to make it function more effectively, many of the reform measures promoted by the World Bank and the IMF in Tanzania and elsewhere in Sub-Saharan Africa during the 1980s constituted just as much of a threat to the survival of this sector as the various challenges that the political stratum had itself mounted in previous periods. 'Much of the opposition to privatization in Tanzania', for example, has resulted from the fact that 'it would threaten the access to privilege and resources that has been so important to many state officials. After all, top parastatal officials have enjoyed certain privileges available only to a few in Tanzania.'[63]

Reinforced as it was by an enduring ideological aversion towards increased private ownership (especially in certain strategic sectors), the political opposition thus exerted on behalf of the public sector was sufficient to prevent privatization of the vast majority of public enterprises in the industrial sector (though some direct sales to, or joint-venture agreements with private investors have occurred).[64] For what one suspects are much the same reasons, moreover, by the end of the 1980s not even the most unprofitable of those enterprises had been forced to close down.[65] As a result, the absolute number of firms in the public sector has remained largely intact: in 1989, for example, there were still more than four hundred state-owned enterprises in Tanzania, with over a hundred of them in the industrial sector alone.[66]

Nor, it would seem, has the economic and technological behaviour of these industrial enterprises been substantially altered over the period in question, in spite of the fact that a series of macro policy reforms has been successfully implemented (in relation, most notably, to the liberalization of foreign trade, domestic prices and financial markets). With no major change in either the penalties that attached to resource misallocation, or in the rewards that attached to improved economic performance, managers of parastatals apparently saw little need to alter their accustomed patterns of behaviour, which tend, as we have seen, to be dominated by a concern with enterprise size rather than profitability and, relatedly, with the 'taking' of existing traits rather than the 'making' of new ones. (These concerns, we further suggested, tend to manifest themselves in a rather indiscriminate demand for, and an acute dependence on, foreign finance in one form or another.)

During the 1980s, for example, though there was a distinct shift in emphasis from investments in entirely new projects to the rehabilitation of existing capacity, the latter activities, according to Wangwe, exhibit many of the same technological features as the former. In particular,

> Rehabilitation programmes have been driven by the modernisation and output maximisation objectives in the static sense in which consideration of technological learning and building of technological capabilities over time are secondary. The rules of the game here are similar to those observed in the design and implementation of new projects and subsequent capacity expansions . . . the rehabilitation programmes tend to reinforce the same features of dependence and low technological learning as those observed in the pre- and post-investment phases in the industrialisation process in Tanzania.[67]

This observation aside, however, there is disappointingly little case-study material covering the technological behaviour of the public sector during the period of structural adjustment reforms (that is, from 1986 to the mid-1990s). Research designed to fill this gap should include an extension of the political economy analysis from the previous section, for it was clearly shown there that the technological behaviour of the public sector is closely related to the complex interplay between the managers of public enterprises on the one hand and the political leadership on the other.

CONCLUSIONS

This chapter has sought to add a political dimension to the analytical framework used in Part I. In particular we have argued in favour of a model that is based on public choice theory but which can be traced back to early work on the theory of the privately owned firm. What these literatures have in common is an emphasis on objectives that are usually promoted by an increase in the size of an institution.

On the basis of a detailed analysis of those variables – at the level of the public sector as well as at the level of certain specific institutions – we find that much of what seems paradoxical about the technological behaviour of public enterprises in Sub-Saharan Africa can be explained with reference to the benefits that accrue to large rather than small enterprises.[68] From a policy point of view, what this conclusion suggests is that even with successful macro policy reform, the inducement of change in existing patterns of technological behaviour in the public sector will be far from easy. For, as we have described it in this chapter, the problem is not simply that alternative, more appropriate forms of behaviour currently go largely unrewarded; it is also that substantial appeal already attaches to the *existing* state of affairs. All too frequently, however, because it is implicitly based on the former conception of the problem the design of policy tends to underestimate the extent of the reform that is actually required.[69]

What should also be borne in mind in any reform process is that the incentives needed to overcome some of the technological problems described in earlier chapters, may not be the same as the incentives required to redress some of the other problems identified in those chapters. What is needed in order to promote static efficiency, for instance, may not at all be the same as the inducements that are likely to foster the (longer-run) acquisition of indigenous technological capabilities (in much the same way that policies designed to promote a purely static notion of comparative advantage may often diverge from what more dynamic considerations would dictate).

It is instructive from this point of view to consider the performance-evaluation system that was introduced in Pakistan in 1988, as part of an effort to link public enterprise performance with payments to its management and employees.[70] The system took the form of bonus payments to firms in the industrial sector whose performance conformed to certain pre-assigned targets (in the most successful enterprises, for example, the bonus amounted to three months' salary). Because there is some reason to think that this system may have led to an increase in operating efficiency among at least some firms,[71] the Pakistan experience with financial incentives deserves to be closely studied by other countries where public sector reforms are now under way (as too does the subsequent Korean case which was based on a similar system of financial incentives but which relied more extensively on non-material rewards such as public recognition of and medals for outstandingly good performance).[72] What the Pakistan case also illustrates, however, is the danger of focusing exclusively on short-run improvements in performance, since public enterprises then tend to 'sacrifice expenditures with short-run costs and only long-run benefits'.[73] The bias thus exerted against projects whose benefits accrue only in the relatively long term becomes all the more pronounced when (as is so frequently the case in developing countries), decision-makers in the public sector hold office for no more than a few years. And the more pronounced is this bias, in its turn, the more will short-term gains tend to be achieved at the expense of technologically-oriented activities – such as training, maintenance and research – whose benefits tend to be realized only after a longer (and in some cases a much longer) period of time.

One way of tackling this problem would be to include performance targets to serve as proxies for the acquisition of indigenous technological capabilities of various kinds. (For this purpose, one could usefully draw on the growing literature concerned with the conceptualization and measurement of these variables at the level of the individual enterprise.[74]) Another way of inducing managers in the public sector to pay attention to relatively long-run technological and other economic issues would be to grant deferred rather than immediate awards.[75] In one variant of this idea, for example, managers would be paid a salary during their tenure in office, but would subsequently (after a period of say five years) be eligible for a financial award if the benefits from investments in technological capability-building were to materialize over this extended period.

The extent to which these and other changes in incentives actually give rise to behavioural modifications in the public sector will depend partly, however, on whether complementary reforms are introduced in connection with foreign aid transfers to Sub-Saharan Africa. For, as Chapter 6 was at pains to emphasize, the incentives that govern those transfers tend to produce just as much of a technological bias as the prevailing inducements in the public sector of the recipient countries themselves. By reducing the extent to which aid is regarded as 'abundant' in donor institutions, reforms of the kind suggested in Chapter 6 would reinforce changes in the same direction that are needed within recipient country institutions as well.[76]

Notes

1. Mueller (1989), pp. 1–2.
2. Ibid., p. 247.
3. Grindle (1991).
4. Ibid., p. 52.
5. As Grindle sees it, the public choice model 'is least applicable to the dynamics of policy making in developing countries when it takes a society-centered approach, that is, when it is based on assumptions about interest mobilization and more or less acquiescent government response to lobbying activities. It is more applicable when it replaces this society-centric view with a more state-centric perspective based on political elites who are actively engaged in maximizing their political power or on rent-seeking bureaucrats'. Ibid., p. 51.
6. Ibid., p. 55.
7. Gallagher (1991).
8. Bates (1981).
9. Baumol (1967), p. 87. Baumol also believed, however, that the typical oligopolistic firm would be subject to a minimum profit constraint.
10. Williamson (1964).
11. Niskanen (1973), p. 8, emphasis in original. Niskanen's approach has drawn criticism from numerous sources, including Mueller (1989), Dunleavy (1991), Jackson (1982), Self (1985) and Breton and Wintrobe (1975).
12. Niskanen (1973), p. 22.
13. Ibid., p. 25.
14. Our discussion of these mechanisms is based on Dunleavy (1991).
15. Ibid., p. 155.
16. Ibid.
17. Mueller (1989), p. 251, emphasis in original.
18. De Alessi (1969).
19. Niskanen (1975), p. 639.
20. Jackson (1982), p. 202.
21. The growth of parastatals in Tanzania is discussed in Mukandala (1988) and Clark (1978).
22. Swanson and Wolde-Semait (1989).
23. Ibid.
24. Williams (1987), p. 48. He continues:

    > Government and bureaucracy in Africa have grown, in part, to tackle the problems of independence and development, but the size and appetite of the bureaucratic state has become, in some cases, a major problem in itself. The creation and expansion of the bureaucratic state in Africa has engendered and exacerbated gross inequalities and disparities.

    It is not only in relation to Africa that the dominance of the bureaucracy has been claimed. See for example Pedersen (1992) on the Indian case.
25. Chazan et al. (1988), pp. 52–3, emphasis added.
26. Horn (1988).

27. Self (1985), p. 140. 'Weber viewed bureaucracy as a highly efficient instrument for applying public policies with precision and consistency. Its "rationality" consisted in the application of precise techniques of legal standards and financial accounting to the work of government, which was to be achieved organizationally through a hierarchy of offices and the detachment of bureaucrats from any personal interest in the results of their decisions', ibid., p. 139. Thus defined, the Weberian ideal has a distinct affinity with the way in which the state was depicted in the early writings of development economists, namely, as an instrument through which social welfare is maximized.
28. Leonard (1987), p. 901.
29. See for example the Ndegwa Commission Report in Kenya (1982).
30. R. Williams (1987), p. 46. See also Sandbrook (1985), Hyden (1983), Young (1986), and Abernethy (1988).
31. Abernethy (1988), p. 191.
32. Ibid., p. 192. See also Young (1986).
33. Ibid., p. 195.
34. Ibid., p. 196.
35. Ibid., emphasis in original.
36. Ibid., p. 198, emphasis in original.
37. Ibid., pp. 203–4, emphasis in original. On this point see also Williams (1987).
38. See for example Leonard (1987), Lancaster (1988), Williams (1987), Sandbrook (1985), Hyden (1983), Young (1986), Abernethy (1988), Chazan et al. (1988).
39. The summary argument here follows Sandbrook (1985).
40. Sandbrook (1985), p. 39.
41. See Hyden (1983).
42. Sandbrook (1985).
43. Hyden (1983), p. 17.
44. Bates (1981).
45. Bates (1988), p. 352.
46. According to Ayub and Hegstad (1986), inefficiencies in the public sector of many developing countries were 'tolerated and even ignored' as long as industrial growth rates were rapid and resources were abundant. With the reduced growth rates and resource constraints of recent years, however, has come a recognition that public enterprises can seriously impede macroeconomic performance.
47. Wilson (1978).
48. Ibid., p. 302.
49. Ibid., pp. 343–4.
50. Ibid., p. 296.
51. Ibid.
52. Mukandala (1988).
53. Ibid., p. 29.
54. Ibid. Here again the relative freedom of the public enterprise, as against other types of institutions in the public sector, bears emphasis. Indeed, according to Mukandala 'The operational environment for parastatals in Tanzania was a dream come true for chief executives at both the holding

corporation and subsidiary company levels. The institution of the holding system created a buffer between subsidiary companies and the state and loosened the ties between them', ibid., p. 115.

55. Ibid, p. 32. There were reportedly also attempts on the part of the managerial stratum to counter the attacks made by the political stratum on managerial incomes. For example, 'They engineered an ingenious shift in emoluments criteria from organizational effectiveness and efficiency towards size and strategic nature of functions – ingenious because every organization ranging from the central bank through mortuaries to chicken farms could marshal arguments to show that their functions were *unique*, and *critical to national survival*', ibid., p. 133, emphasis in original.
56. Ibid., p. 327.
57. Ibid., p. 333.
58. Ibid., p. 313, emphasis added.
59. Ibid., p. 129.
60. Ibid., pp. 128–9.
61. Another manifestation of this lay in the fact that 'research, investigations and feasibility studies were', in Mukandala's words, 'watered down', ibid., p. 334.
62. More generally, 'The management of projects under implementation . . . was poor', ibid.
63. Callaghy and Wilson (1988), p. 194. In Kenya, too, privatization has encountered formidable political and other problems. Indeed, by 1993 only one state enterprise had actually been sold in that country (*The Economist*, 1 May 1993). In certain other African countries, however, privatization has been more successful, as the survey by Callaghy and Wilson (1988) clearly demonstrates. See also Cook and Fitzpatrick (1988). The most recent data prepared by the World Bank (1994) show that in most African countries, privatization has been very limited.
64. Several examples are cited in Due (1993) who also notes that 'Privatization deals have taken much longer than anticipated in Tanzania' (p. 1984).
65. World Bank (1991).
66. Ibid.
67. Wangwe (1992), pp. 290–1. Problems with rehabilitation programmes are described also in the World Bank (1994). Macro-economic policy reforms do seem, however, to have brought about a desirable alteration in the extent to which firms of varying degrees of efficiency contribute to total output in the industrial sector. In particular, the output of relatively inefficient firms has contracted, whereas that of smaller, more labour-intensive enterprises has expanded (World Bank, 1991).
68. This should of course be interpreted as a central tendency among firms in the public sector rather than as a behavioural rule that admits of no exceptions.
69. Policy that is based entirely on a narrowly economic diagnosis of what is required, for example, will tend to fall into this category because of its misconception of the factors that underlie the existing state of affairs. A good recent discussion of this point in relation to economic liberalization in Africa is to be found in Sandbrook (1993).
70. For a detailed description of this case see Shirley (1991). As suggested by

Knight (1992), the Chinese experience with incentives in the public sector also warrants attention.

71. Shirley (1991) has undertaken a fairly detailed empirical analysis of this question.
72. Ibid. There is of course also much to be gained from a review of the experience with performance-related pay systems in the *private* sector. Indeed, the majority of privately owned firms in the United States and other developed countries now use financial incentives of one kind or another to motivate their workers (*The Economist*, 29 January 1994).
73. Shirley (1991), p. 264.
74. See for example Lall (1992).
75. This idea is advanced by Niskanen (1973), as part of a detailed discussion of how improved incentive systems for bureaucrats might be designed.
76. Mytelka (1985), for example, describes the case of 'Kiltex', a state-owned textile enterprise in Tanzania, which apparently used the problems created by a scarcity of foreign exchange to make numerous technological adaptations.

References

D. Abernethy (1988). 'Bureaucratic Growth and Economic Stagnation in Sub-Saharan Africa', in S. Commins (ed.), *Africa's Development Challenges and the World Bank*, Lynne Rienner.

L. de Alessi (1969). 'Implications of Property Rights for Government Investment Choices', *American Economic Review*, vol. LIX(1).

M. Ayub and S. Hegstad (1986). 'Public Industrial Enterprises: Determinants of Performance', Industry and Finance Series, vol. 17, World Bank.

R. Bates (1981). *Markets and States in Tropical Africa*, University of California Press.

R. Bates (1988), 'Governments and Agricultural Markets in Africa', in R. Bates (ed.), *Toward a Political Economy of Development: A Rational Choice Perspective*, University of California Press.

W. Baumol (1967). *Business Behavior, Value and Growth*, 2nd edn., Harcourt, Brace and World Inc.

D. Boes (1989). *Public Enterprise Economics*, North-Holland.

A. Breton and R. Wintrobe (1975). 'The Equilibrium Size of a Budget-Maximizing Bureau: A Note on Niskanen's Theory of Bureaucracy', *Journal of Political Economy*, vol. 83(11).

T. Callaghy and E. Wilson (1988). 'Africa: Policy, Reality or Ritual?', in R. Vernon (ed.), *The Promise of Privatization: A Challenge for US Policy*, Council on Foreign Relations.

N. Chazan, R. Mortimer, J. Ravenhill and D. Rothchild (eds) (1988). *Politics and Society in Contemporary Africa*, Lynne Rienner.

W.E. Clark (1978). *Socialist Development and Public Investment in Tanzania*, University of Toronto Press.

P. Cook and C. Fitzpatrick (eds) (1988). *Privatization in Less Developed Countries*, Harvester Wheatsheaf.

J. Due (1993). 'Liberalization and Privatization in Tanzania and Zambia', *World Development*, vol. 21(12).
J. Dunleavy (1991). *Democracy, Bureaucracy and Public Choice*, Harvester Wheatsheaf.
M. Gallagher (1991). *Rent Seeking and Economic Growth in Africa*, Westview.
M. Grindle (1991). 'The New Political Economy: Positive Economics and Negative Politics', in G. Meier (ed.), *Politics and Policy Making in Developing Countries: Perspectives on the New Political Economy*, ICS Press, an International Center for Economic Growth Publication.
P. Heller and A. Tait (1984). 'Government Employment and Pay: Some International Comparisons', IMF Occasional Paper no. 62, International Monetary Fund, Washington, DC.
M. Horn (1988). *The Political Economy of Public Administration: Organization, Control and Performance of the Public Sector*, PhD thesis, Harvard University.
G. Hyden (1983). *No Shortcuts to Progress*, Heinemann.
P. Jackson (1982). *The Political Economy of Bureaucracy*, Phillip Allan.
J. Knight (1992). 'Public Enterprises and Industrialization in Africa', in F. Stewart, S. Lall and S. Wangwe (eds), *Alternative Development Strategies in SubSaharan Africa*, Macmillan.
S. Lall (1992). 'Technological Capabilities and Industrialization', *World Development*, vol. 20(2).
C. Lancaster (1988). 'Political Economists and Policy Reformers in Africa', Economic Development Institute Working Paper, World Bank.
D. Leonard (1987). 'The Political Realities of African Management', *World Development*, vol. 15(7).
D. Mueller (1989). *Public Choice II: A Revised Edition*, Cambridge University Press.
R. Mukandala (1988). *The Political Economy of Parastatal Enterprise in Tanzania and Botswana*, PhD thesis, University of California, Berkeley.
L. Mytelka (1985). 'Stimulating Effective Technology Transfer: the Case of Textiles in Africa', in N. Rosenberg and C. Frischtak (eds), *International Technology Transfer: Concepts, Measures and Comparisons*, Praeger.
P. Ndegwa (1982). *Working Party on Government Expenditures*. Report and Recommendations of the Committee appointed by His Excellency the President, Nairobi, Kenya.
J. Nellis (1986). 'Public Enterprises in Sub-Saharan Africa', World Bank Discussion Papers.
W. Niskanen (1973). *Bureaucracy: Servant or Master?* Institute of Economic Affairs, London.
W. Niskanen (1975). 'Bureaucrats and Politicians', *Journal of Law and Economics*, vol. 18(4).
J. Pedersen (1992). 'State, Bureaucracy and Change in India', *Journal of Development Studies*, vol. 28(4).
R. Sandbrook with J. Barker (1985). *The Politics of Africa's Economic Stagnation*, Cambridge University Press.
R. Sandbrook (1993). *The Politics of Africa's Economic Recovery*, Cambridge University Press.
P. Self (1985). *Political Theories of Modern Government*, Allen & Unwin.
M. Shirley (1991). 'Evaluating the Performance of State-Owned Enterprises in

Pakistan', in R. Ramamurti and R. Vernon (eds), *Privatization and Control of State-Owned Enterprises*, Economic Development Institute, World Bank.
D. Swanson and T. Wolde-Semait (1989). 'Africa's Public Enterprise Sector and Evidence of Reforms', World Bank Technical Paper no. 95.
S. Wangwe (1992). 'Building Indigenous Technological Capacity: A Study of Selected Industries in Tanzania', in F. Stewart, S. Lall and S. Wangwe (eds), *Alternative Development Strategies in SubSaharan Africa*, Macmillan.
R. Williams (1978). *Political Corruption in Africa*, Gower.
O. Williamson (1964). *The Economics of Discretionary Behavior*, Prentice Hall.
E. Wilson (1978). *The Political Economy of Public Corporations in the Energy Sectors of Nigeria and Zaire*, PhD thesis, University of California, Berkeley.
World Bank (1986). *Financing Adjustment with Growth in Sub-Saharan Africa, 1986–1990*, Washington, DC.
World Bank (1991). *Tanzania Economic Report: Towards Sustainable Development in the 1990's*, Report no. 9352–TA, Washington, DC.
World Bank (1994). *Adjustment in Africa: Reforms, Results and the Road Ahead*, A World Bank Research Policy Report, Oxford University Press.
C. Young (1986). 'Africa's Colonial Legacy', in J. Whitaker and R. Berg (eds), *Strategies for African Development*, University of California Press.

8 The Centralized African State and its Implications for Technological Behaviour in the Public Sector

Though the foregoing political economy analysis does allude to certain distinctive aspects of the African state, it does not however incorporate the important fact that, by the standards of most other developing countries, this state is typically run along highly centralized lines. Such a comparison invites a number of questions, of which the following are the most essential for our purposes: Through what mechanisms and in which directions has the pronounced centralization of state power in Africa influenced the pattern of technological behaviour in the public sector? How amenable to change are the influences thus exerted? In addressing these questions, the argument of this chapter will be that pronounced concentration of power at the central level constitutes yet another reason for the stylized patterns of technological behaviour described in Chapters 1 and 3. And like the other political factors which were described in the previous chapter, the centralization of state power has a powerful self-perpetuating quality which has tended to undermine policy reforms in the 1970s and 1980s and which continues to pose a fundamental challenge to contemporary political and economic change in Africa.

EXTREME CENTRALIZATION OF THE AFRICAN STATE

'In comparison to other regions of the Third World,' writes Stren, 'sub-Saharan Africa has typically paid little attention to what we in the west call "local government".'[1] The data that he cites in support of this contention are shown in Table 8.1, which indicates the differential extent to which total (non-agricultural) employment in the public sector of three developing regions is made up of employment in central rather than local institutions.

In Asia, for example, of the 21.9 per cent of total employment outside agriculture that the government contributes, 63 per cent takes place in central institutions and the remaining 37 per cent occurs at the local level. In Africa, by contrast, only 6 per cent of total government employment outside agriculture is accounted for outside the central agencies of the

Table 8.1 Share of government in non-agricultural employment, by developing region

	Africa	*Asia*	*Latin America*
Central government	30.8 (94%)	13.9 (63%)	15.8 (79%)
Local government	2.1 (6%)	8.0 (37%)	4.2 (21%)
Total government	32.9	21.9	20.0

Note: Based on a survey covering the period between late 1970s and early 1980s. The percentages in parentheses refer to columns.
Source: *World Development Report*, 1983, World Bank, p. 102.

state. All in all, 'While these figures are only indicative of relative differences, the impression they give of extreme centralisation in Africa is nevertheless correct'.[2]

It is not that countries in the region failed to recognize the many problems that attend so pronounced a degree of centralization, nor that no attempts have been made to attack the source of those problems. On the contrary, during the 1970s and 1980s decentralization came to be regarded as an important policy goal and in some African countries – such as Kenya, Tanzania and the Sudan – policies to achieve this goal were widely promulgated.[3] The problem, rather, was that most such efforts appear to have achieved only a very limited degree of success. This conclusion emerges compellingly from a review of decentralization policy outcomes in the four countries listed in Table 8.2, and other assessments of the African experience with decentralization come to very much the same conclusion.[4]

Among the many factors that have contributed to these disappointing outcomes,[5] the most noteworthy from the political economy point of view of the previous chapter, is the role played by those who, as members of central government institutions, stood to lose perhaps the most from the policy reforms. In his comparison of Kenya, Tanzania and the Sudan, Rondinelli, for example, refers to the 'continuing resistance of central government officials to "decision-making from below"'.[6] Just as the 'managerial stratum' in Tanzania was shown in the previous chapter to have thwarted many of the attempts to attenuate its power, so too did the central bureaucracies in Kenya, Tanzania and the Sudan seek by one means or another, to undermine the decentralization policies that were widely promulgated in each of those countries.

In Kenya and Tanzania for example,

Table 8.2 Outcomes of decentralization policies: selected countries in Sub-Saharan Africa

Kenya	'decentralized structures in Kenya are but extension of the centre, carrying tasks assigned to them through the regular ministries. Even the freedom of the various self-help projects is subject to state intervention and control.' (p. 83)
Tanzania	'Despite all the hectic reforms and changes, the authority of the centre is still unchallenged in terms of resource allocation and political control of both urban and district councils. The Tanzanian central government has not been able to do away with its centralization tendencies.' (p. 85)
Zambia	'To all intents and purposes, the new system merely reinforced central control rather than promoting decentralization and grassroots participation in the decision-making process.' (p. 86)
Nigeria	'Despite the wide-ranging reforms that were introduced over the years, government officials at the state government level continued to treat local authorities as subordinate organizations, which deserved to be directed but not listened to, very much reflecting past relationships and attitudes. The ideals of local government . . . proved to be irrelevant.' (p. 89)

Source: Asmerom (1992).

> the central bureaucracies successfully maintained control over regional, provincial, and district development planning; when they were not able to obstruct the transfer of functions, they attempted to extend their control to lower levels of administration. In the Sudan, the only effective force for making decentralization work was within the office of the provincial commissioner, but bureaucrats within the commissioners' offices were often ministry technicians and neither they nor provincial executive councils were willing to extend authority for decision making to local councils. . . .
>
> The ability of the central bureaucracy to penetrate local administrative units or reconsolidate power was evident in all three countries.[7]

Despite their dissimilar ideologies, what Kenya and Tanzania appear to have in common therefore is a relatively large, centralized bureaucracy, that has proved to be remarkably resilient in the face of strong countervailing forces of various kinds.[8] This similarity is important for our purposes because of the influence that centralization exerts on the technological behaviour of the public sector and hence the further light that can thereby be thrown on the comparison between these two countries that was undertaken in Chapter 3.

TECHNOLOGICAL IMPLICATIONS OF THE CENTRALIZED STATE

The balance between national and local institutions in a particular country would not influence the technological behaviour of the public sector as a whole if the institutions happened to be associated with similar technologies, for then it would not matter if the overall institutional mix happened to be skewed in one direction or another. If this is not the case, however, then the differential amounts of resources controlled by national rather than local institutions will indeed tend to skew the overall technological outcome in a particular direction.[9]

In much of Africa, we suggest, it is precisely this type of institutional imbalance that gives rise to and is indeed reinforced by, the prevalence of inappropriate technology and underdeveloped technological capabilities in the public sector. In particular, it is the dominance of central institutions that skews the overall technological balance in the public sector towards large-scale capital and import-intensive technologies rather than the smaller-scale, labour-intensive alternatives that decentralized institutions would be more likely to select.[10] In contrast perhaps most strikingly with China, where, partly on account of a relatively decentralized system, a particularly balanced mix of technologies is to be found,[11] what one usually finds in Africa is a remarkably unbalanced and inappropriate pattern of technological behaviour in the public sector. It matters little from this point of view, whether the state is ideologically inclined towards capitalism (as in Kenya) or towards socialism (as in Tanzania). Indeed, it is precisely the absence of a systematic correlation between ideology and the institutional distribution of state power, that helps to explain why state-owned enterprises in countries as diverse as Kenya and Tanzania exhibit such similar forms of technological behaviour.

In addition to the influence thus exerted on the balance of technology in the public sector as a whole, the degree to which public sector institutions are run on centralized lines also affects the extent to which this balance can *itself* be altered. The reason for this has mainly to do with the fact that the production of a given level of output by labour, as opposed to capital-intensive technologies, is not normally associated with the same number of plants. On the contrary, use of the former technologies generally implies a larger (and in some cases a much larger) number of enterprises than does use of

Table 8.3 Numbers of plants required under alternative methods of producing a given increase in output in selected manufacturing industries.

Sector	*Increment in output*	*Number of plants*
Cashew processing (Tanzania)	425,000 tons *	
Mechanised		85
Semi-mechanised		340
Textiles (Tanzania)	282 million m^2 *	
Capital-intensive		10
Labour-intensive		10
Footwear (Tanzania)	28 million pairs *	
Large-scale		3
Small-scale		426
Wood products (Tanzania)	516,000 m^3 *	
Capital-intensive		96
Labour-intensive		652
Soap (Tanzania)	54,000 tons *	
Large-scale		26
Small-scale		627
Bread-baking (Tanzania)		
Automated method	100,000 loaves	1
Semi-automated method		10
Labour-intensive		50
Sugar (Kenya)	380,000 tons	
Vacuum-pan method		2.6
Open-pan method		
100 tons of cane		162
200 tons of cane		81
Bricks (Botswana)	20–25 million	
Capital-intensive		1
Medium-scale		5
Labour-intensive		11

Sources: For all cases other than bricks, sugar and bread, Roemer et al. (1976). For bread-baking, Coulson (1982) and Green (1978); for sugar, Kaplinsky (1987); and for bricks Kaplinsky (1990).

* Incremental output data refer to the period between 1975 and 1995 and they form part of the strategic planning exercise in Tanzanian industry, as formulated by Roemer et al. (1976). The other incremental output data are based on the individual case-studies mentioned above.

the latter, as Table 8.3 clearly illustrates for a range of different manufacturing activities.

The first five sectoral comparisons in the table are drawn from a detailed planning exercise conducted by Roemer, Tidrick and Williams, which sought to present to policy-makers the strategic choices in industry that were thought to have confronted Tanzania over the 20-year period from 1975 to 1995.[12] In the case of cashew processing, for example, output was projected to increase by 425,000 units over the period, and this increase could have been achieved on the basis of either mechanized or semi-mechanized production techniques. In the former case 85 additional plants would have been required, as against the much larger figure of 340 that the latter technology would have entailed. With the exception of textiles, the other sectors exhibit similar and indeed even more pronounced differentials between the more and less capital-intensive alternatives. In Kenya, for example, an annual sugar output of 380,000 tons could be produced either in 2.6 plants using vacuum-pan methods, or in 81–162 small-scale units using the alternative open-pan technology. In the former (vacuum-pan) case, the numbers employed would be 4420, whereas in the case of the smaller-scale open-pan technologies the numbers employed would range from 25,920 to 29,160 depending on whether the crushing capacity was equal to 100 or 200 tons of cane per day.[13]

The incremental units of production that, in these and the other sectors shown in Table 8.3, are associated with relatively labour-intensive technologies, in turn, presuppose a relatively decentralized institutional framework, as had already been recognized in relation to the Third Five Year Plan in Tanzania. 'The degree to which small-scale labour-intensive techniques are possible', wrote David Phillips at the time:

> depends on the degree of decentralisation of control and ownership. The Tanzanian economy is characterised by centralised control, strongly influenced by external trade and foreign technology and capital dependence. It follows that Tanzania's objective of relocation of industry and small-scale industrial development will not be achieved, and technological alternatives will not be adopted, unless a far-reaching redistribution of control occurs in industry, involving local enterprises, cooperatives, district development corporations and other organisations able to control appropriation and distribution of investible surpluses, bank credit, savings, employment policy, raw material supplies and markets.[14]

Although this line of argument makes the choice of technology an unavoidably political issue, as Phillips rightly stresses, it can also be described in the (micro-economic) terms of a 'third factor of production', which, in the form of, say, a decentralized organizational capability in the public sector, is highly complementary to the more familiar inputs of capital and labour. Where this 'third factor' is absent or extremely scarce, production based on labour-intensive technology will thus tend to be inefficient or perhaps even entirely impossible.[15] Put still another way, the point is that: 'The opportunity cost of initiating the fertilizer plant employing 400 workers may in fact be a smaller project, in terms of capital, which would still employ only 400 workers, not two or three smaller projects employing 1000 workers.'[16]

It is this type of reasoning that helps to account for what, to many observers of the controversial automated bakery project in Tanzania, seemed like highly inappropriate technology. In this case, to which we have already referred in a previous chapter, a small-scale alternative to the automatic bakery clearly existed.[17] This alternative not only entailed a substantially lower capital cost but also reduced import costs in comparison with the large-scale bakery.[18] Furthermore, the spare parts and replacement machinery associated with the small-scale alternative were of a kind that could have been locally produced and which could therefore have assisted in the acquisition of technological capabilities in the domestic capital goods sector.

Yet in spite of all these advantages small bakeries were rejected in favour of what, from the point of view of most important national goals, seemed like a distinctly inferior alternative. 'The real difficulty', according to Green, 'was that there was no evident decentralized public sector institution to operate small bakeries.'[19] Nor, apparently, was any process set in train that would appreciably have redressed this rather glaring institutional weakness. Several years after the project had begun, for example, there were still 'no guidelines as to what degree of risk, of project cost and of institutional weakness are acceptable if they contribute to decentralized public sector operations and to the use of simpler, less capital-intensive technologies'.[20] A similar point emerges from Mytelka's interesting study of four manufacturing industries in Côte d'Ivoire, which indicates that in spite of the all too obvious failing of most large-scale public sector projects in those sectors, 'little has been done to encourage alternatives'.[21] More likely than not, therefore, *future* decisions would *again* (in Hirschman's appealing and by now familiar terminology) take the form of 'trait-taking' rather than 'trait-making' and would thereby justify once more the selection of manifestly inappropriate technologies in the public sector.[22]

At each such moment of time, that is to say, a purely *static* comparison would indeed favour the option that economizes on scarce organizational capabilities at a decentralized level (to the extent that a comparison is made at all; in Kaplinsky's study of brickmaking in Botswana a larger number of small-scale plants was not even considered as an alternative to the large-scale option,[23] and the same goes for Wangwe's account of the cement industry in Tanzania and Holtham and Hazlewood's description of the Mumias sugar project in Kenya[24]). The problem with this sort of comparison even when it is made, however, is that it lacks an explicit assessment of the social costs and benefits of policies designed precisely to *satisfy* the organizational demands associated with small-scale technologies. It lacks, in short, the dynamic perspective that an effective challenge to the status quo seems to require. For, in the absence of such a perspective, what often tends to be overlooked is the possibility that policies designed to *overcome* the existing organizational limitations might deliver an amount of benefits to society that would exceed their social costs.[25]

Added, as this limiting tendency mostly was, to a no less static, 'trait-taking' type of view of the availability of local skills and technologies for large-scale investment projects in industry,[26] it is hardly surprising that indigenous technological capabilities in the public sector as a whole progressed so little in Sub-Saharan Africa during the 1970s and 1980s.

CONCLUSIONS

What we have described in this brief chapter, then, is a set of mutually reinforcing relationships that can be summarized as follows. The predominance of central rather than local government institutions in Africa typically creates a marked imbalance in the overall pattern of technological behaviour in the public sector and a strong bias in favour of certain types of technology. These tendencies, in turn, reinforce the dominance of central institutions and make it all the more difficult to introduce smaller-scale technologies, which, because of the much larger number of plants that they usually entail, necessitate an appreciable strengthening of local institutions of various kinds. It is thus not at all surprising that attempts to challenge the political foundations of these cumulative patterns of causality have been strongly resisted, and it may be, as Sandbrook has recently suggested, that in countries ruled until recently by 'authoritarian governments and weak civil societies', the process of 'strengthening and democratizing people's and voluntary associations will take years'.[27]

Yet even now the continent is not entirely devoid of bright spots. Uganda, for example, seems to be in the midst of an interesting experiment to replace the centralized party-political system with a decentralized system of local government in which the population votes for individuals rather than parties.[28] In two other countries, moreover, Zimbabwe and Botswana, 'the continued vitality of local government . . . supports the notion that devolution can work effectively in Africa'.[29] Though they do not seem to have been examined from this particular point of view, these cases might help to identify an alternative model of technological behaviour, which, because it is based on decentralized rather than centralized institutions of the state, tends to favour rural over urban areas and clusters of small-scale labour-intensive units of production over a relatively small number of large-scale, capital-intensive plants.

Notes

1. Stren (1988), p. 219.
2. Ibid. As with so many of the features of the African state in the post-independence period, its highly centralized character owes much to the influences that were exerted during the colonial period. Hyden (1983, pp. 111–12) for example, points out that during this period 'Neither participation nor accountability to local structures was given priority'. See also Young (1986), Sandbrook (1985) and Chazan et al. (1988).
3. Rondinelli (1983).
4. See the recent review by Asmerom (1992) for example. See also Sandbrook (1993).
5. Rondinelli (1983) has an extensive discussion of these.
6. Ibid., p. 97.
7. Ibid., pp. 97–8. Examples can also be readily cited from the West African experience. In Ghana, for instance, 'those at the center' were apparently 'reluctant to accept any diminution in their decision-making powers' while regional officials sought to 'retain control' over the field officers in their jurisdiction. Cited in Rondinelli and Nellis (1986), p. 12.
8. For Tanzania, Coulson reports on an interesting study which compared small-scale projects in three districts prior to and after the 'decentralization reforms' of 1972. Though these were intended to transfer additional power to local organizations, in the event, additional power accrued to government officials instead. See Coulson (1982), p. 319. In the aggregate, examples such as these help to explain the failure of the well-known Ujamaa Village Scheme in that country.
9. Stewart (1987) uses the general term 'composition of units' to refer to the proportion of resources controlled by different economic agents.
10. In general, according to Stewart, 'small-scale units . . . use more labor-intensive technologies than large-scale units; whether they are public or private. By shifting the control of total resources the government may bring about more appropriate technology in the economy as a whole, even though it has not altered the decisions of any particular unit.' Ibid., p. 6.
11. See for example Perkins (1983).
12. Roemer, Tidrick and Williams (1976).
13. Kaplinsky (1987).

14. Phillips (1979), p. 86.
15. This recognition appears very briefly in the article by Roemer et al., under the guise of higher 'administration costs' that a labour-intensive form of industrialization would probably have entailed.
16. Clark (1978), p. 213.
17. Green (1978).
18. Ibid.
19. Ibid., p. 15.
20. Ibid., p. 20.
21. Mytelka (1992). In the cement industry, for example, a study financed by UNIDO recommended the production of cement block which 'is less import intensive than cement, uses large quantities of cheap laterite clay, can be produced in small decentralised units thereby decreasing the cost of building materials and of investment costs at the same time as it increases employment'. However, 'The scarcity of public funds and the lack of artisans to staff these small enterprises seriously constrain the government's ability to implement this programme. Pressures from the cement industry are likely to contribute to restraining this initiative' (p. 262).
22. Green himself was not very optimistic about the extent to which the lessons from the bakery project would be reflected in future decisions. 'The issue', he wrote, 'is whether Tanzanian bodies will have taken steps to be better able to analyze, propose, discuss, decide and implement. The 1977 Dar butchery case does not give grounds for a very optimistic prognosis as of 1978.' (Green, 1978, p. 25).
23. Kaplinsky (1990).
24. Holtham and Hazlewood (1976), Wangwe (1992). The latter cites a study showing that in Tanzania Portland cement can be produced in small-scale plants at roughly half the cost of production in large-scale plants.
25. The same type of problem occurs, as Pack has rightly emphasized, in relation to other inputs that a labour-intensive development strategy would entail.

    For example, it often is asserted that adoption of labor-intensive techniques is constrained by their requirements for maintenance and skilled manpower and, if used equipment is concerned, by the unavailability of spare parts and the problems of identifying good used machines. . . . Usually such an acknowledgement is transmitted into a counsel of despair: the shortages identified make it *impossible* for appropriate techniques to be adopted. . . . An alternative approach is superior: investment may be required to remove or reduce the alleged shortages, and the cost of the necessary outlays should be compared with the calculated benefits. See Pack (1982), p. 21, emphasis in original.

26. As noted in Chapter 1 above.
27. Sandbrook (1993), p. 146. See also Herbst (1990). Both these authors deal at length with the political implications and problems associated with structural adjustment policies in Sub-Saharan Africa. In that they challenge the centralization of state power, those policies tend to encounter some of the same political problems as a move towards decentralized, small-scale technologies in the public sector.

28. Review of *The Search for Africa*, by B. Davidson, in *The Economist*, 23 April 1994.
29. Sandbrook (1993), p. 72.

References

H.K. Asmerom (1992). 'Decentralization and Administrative Capacity Building at Regional and Local Levels: With some Examples from Sub-Saharan Africa', in J. Heins (ed.) *Falende Overheden? Bestuur, Politiek en Besluitvorming in de Derde Wereld*, VU Uitgeverij, Amsterdam.

N. Chazan, R. Mortimer, J. Ravenhill and D. Rothchild (1988). *Politics and Society in Contemporary Africa*, Lynne Rienner.

W.E. Clark (1978). *Socialist Development and Public Investment in Tanzania*, University of Toronto Press.

A. Coulson (1982). *Tanzania: A Political Economy*, Oxford University Press.

R. Green (1978). 'The Automated Bakery: A Study of Decision Taking Goals, Processes and Problems in Tanzania', Institute of Development Studies, Sussex, Discussion Paper no. 141.

J. Herbst (1990). 'The Structural Adjustment of Politics in Africa', *World Development*, vol. 18(7).

G. Holtham and A. Hazlewood (1976). *Aid and Inequality in Kenya*, Croom Helm.

G. Hyden (1983). *No Shortcuts to Progress*, Heinemann.

R. Kaplinsky (1987). 'Appropriate Technology in Sugar Manufacturing', in F. Stewart (ed.), *Macro-Policies for Appropriate Technology in Developing Countries*, Westview.

R. Kaplinsky (1990). *The Economies of Small*, IT Publications.

L. Mytelka (1992). 'Ivorian Industry at the Cross-Roads', in F. Stewart, S. Lall, and S. Wangwe (eds), *Alternative Development Strategies in SubSaharan Africa*, Macmillan.

H. Pack (1982). 'Aggregate Implications of Factor Substitution in Industrial Processes', *Journal of Development Economics*, vol. 11(1).

F. Perkins (1983). 'Technology Choice, Industrialization and Development: the Case of Tanzania', *Journal of Development Studies*, no. 19.

D. Phillips (1979). 'Industrialization in Tanzania', in K. Kim, R. Mabele and M. Schultheis (eds), *Papers on the Political Economy of Tanzania*, Heinemann.

M. Roemer, G. Tidrick and D. Williams (1976). 'The Range of Strategic Choice in Tanzanian Industry', *Journal of Development Economics*, vol. 3.

D. Rondinelli (1983). 'Decentralization of Development Administration in East Africa', in G.S. Cheema and D.A. Rondinelli (eds), *Decentralization and Development*, Sage.

D. Rondinelli and J. Nellis (1986). 'Assessing Decentralization Policies in Developing Countries: The Case for Cautious Optimism', *Development Policy Review*, vol. 4.

R. Sandbrook (1985). *The Politics of Africa's Economic Stagnation*, Cambridge University Press.

R. Sandbrook (1993). *The Politics of Africa's Economic Recovery*, Cambridge University Press.

F. Stewart (ed.) (1987). *Macro-Policies for Appropriate Technology in Developing Countries*, Westview.

R. Stren (1988). 'Urban Services in Africa: Public Management or Privatization?', in P. Cook and C. Kirkpatrick (eds), *Privatisation in Less Developed Countries*, Harvester Wheatsheaf.

S. Wangwe (1992). 'Building Indigenous Technological Capacity: A Study of Selected Industries in Tanzania', in F. Stewart, S. Lall and S. Wangwe (eds), *Alternative Development Strategies in SubSaharan Africa*, Macmillan.

C. Young (1986). 'Africa's Colonial Legacy', in R. Berg and J. Whitaker (eds), *Strategies for African Development*, University of California Press.

Index